AF342179

Safety Issues Associated with Plutonium Involvement in the Nuclear Fuel Cycle

NATO Science Series

*A Series presenting the results of activities sponsored by the NATO Science Committee.
The Series is published by IOS Press and Kluwer Academic Publishers, in conjunction
with the NATO Scientific Affairs Division.*

General Sub-Series

A. Life Sciences	IOS Press
B. Physics	Kluwer Academic Publishers
C. Mathematical and Physical Sciences	Kluwer Academic Publishers
D. Behavioural and Social Sciences	Kluwer Academic Publishers
E. Applied Sciences	Kluwer Academic Publishers
F. Computer and Systems Sciences	IOS Press

Partnership Sub-Series

1. Disarmament Technologies	Kluwer Academic Publishers
2. Environmental Security	Kluwer Academic Publishers
3. High Technology	Kluwer Academic Publishers
4. Science and Technology Policy	IOS Press
5. Computer Networking	IOS Press

*The Partnership Sub-Series incorporates activities undertaken in collaboration with NATO's
Partners in the Euro-Atlantic Partnership Council – countries of the CIS and Central and Eastern
Europe – in Priority Areas of concern to those countries.*

NATO-PCO-DATA BASE

The NATO Science Series continues the series of books published formerly in the NATO ASI
Series. An electronic index to the NATO ASI Series provides full bibliographical references (with
keywords and/or abstracts) to more than 50000 contributions from international scientists published
in all sections of the NATO ASI Series.
Access to the NATO-PCO-DATA BASE is possible via CD-ROM "NATO-PCO-DATA BASE" with
user-friendly retrieval software in English, French and German (© WTV GmbH and DATAWARE
Technologies Inc. 1989).

The CD-ROM of the NATO ASI Series can be ordered from: PCO, Overijse, Belgium.

Series 1: Disarmament Technologies – Vol. 23

Safety Issues Associated with Plutonium Involvement in the Nuclear Fuel Cycle

edited by

Theodore A. Parish

Texas A&M University,
College Station, Texas, U.S.A.

Vyacheslav V. Khromov

Moscow Engineering Physics Institute,
Moscow, Russian Federation

and

Igor Carron

Texas A&M University,
College Station, Texas, U.S.A.

Kluwer Academic Publishers

Dordrecht / Boston / London

Published in cooperation with NATO Scientific Affairs Division

Proceedings of the NATO Advanced Research Workshop on
Safety Issues Associated with Plutonium Involvement in the Nuclear Fuel Cycle
Moscow, Russia
2–6 September 1997

A C.I.P. Catalogue record for this book is available from the Library of Congress.

ISBN 0-7923-5592-X (HB)

Published by Kluwer Academic Publishers,
P.O. Box 17, 3300 AA Dordrecht, The Netherlands.

Sold and distributed in North, Central and South America
by Kluwer Academic Publishers,
101 Philip Drive, Norwell, MA 02061, U.S.A.

In all other countries, sold and distributed
by Kluwer Academic Publishers,
P.O. Box 322, 3300 AH Dordrecht, The Netherlands.

Printed on acid-free paper

All Rights Reserved
© 1999 Kluwer Academic Publishers
No part of the material protected by this copyright notice may be reproduced or
utilized in any form or by any means, electronic or mechanical, including photo-
copying, recording or by any information storage and retrieval system, without written
permission from the copyright owner.

Printed in the Netherlands

CONTENTS

PLUTONIUM USAGE IN EXISTING/NEAR-TERM REACTORS

FOREWORD

The "VOLGA" conferences, hosted in odd-numbered years by the Department of Theoretical and Experimental Reactor Physics of the Moscow Engineering Physics Institute (MEPhI), are some of the most prestigious technical meetings held in Russia. Traditionally, these conferences present the opportunity for reactor physicists from around the world to gather at MEPhI's holiday camp on the banks of the Volga river (near Tver) to exchange ideas and explore innovative concepts related to nuclear power development. In 1997, NATO became involved in the "VOLGA" meetings for the first time by co-sponsoring "VOLGA97" as an advanced research workshop. This workshop broke with tradition a bit in that the venue was moved from MEPhI's holiday camp to a location nearer Moscow.

The workshop program was effectively organized in order to cover a broad range of topics relating to the theme of the meeting. Generally, the papers concerned safety-related questions associated with utilizing both weapons-grade and reactor-grade plutonium in the nuclear fuel cycle, including facility requirements, licensing issues, proliferation risks, and a variety of advanced concepts for alternative fuel cycles. The program contained a total of ninety-nine papers presented in five days of sessions.

The first plenary session featured four interesting papers which established the international tone of the workshop. The first paper was presented by Bruce Bevard and Alexander Chebeskov of ORNL and IPPE Obninsk, respectively, on the joint US/Russia projects for the disposition of weapons-grade plutonium. Vladimir Kagramanian of the IAEA gave a global perspective on energy needs and the future role of nuclear energy in the context of sustainable development. Alexander Dmitriev of Gosatomnadzor (the Russian Nuclear Safety and Licensing Authority) provided a regulatory-based overview of the options for plutonium disposition in Russia. Finally, David Boyle of Texas A&M University described the activities of the Amarillo National Resource Center for Plutonium in temporary storage, plutonium disposition and public outreach.

In the subsequent sessions, attention was generally given to the role of plutonium and the minor actinides in various types of reactors and in the nuclear fuel cycle. On a practical level, papers from the United Kingdom and France dealt with currently operating facilities for MOX fuel fabrication, including process modifications and safety constraints which would arise in constructing a MOX fabrication plant for weapons-grade plutonium. In contrast, papers from Germany and Russia offered innovative ideas on "denaturing" weapons-grade plutonium with reactor-grade plutonium, requirements for long-term applications of nuclear energy, and the use of new fuel forms to increase proliferation resistance. A feature of the program which was appreciated by the participants was the lively question and answer periods following each paper, and the free expression of divergent viewpoints. These discussions often carried over to the breaks and the social functions as well.

Over 100 participants took part in the workshop. Personnel representing virtually all of the major Russian research institutes and organizations active in nuclear fuel cycle activities, as well as, universities and nuclear power plants attended the workshop. Some of the Russian organizations included the Institute of Physics and Power Engineering-Obninsk, the Kurchatov Institute, the Bochvar Institute of Inorganic Materials, the

Russian Academy of Natural Sciences and the nuclear weapons laboratories at Arzamas-16 and Chelyabinsk-70. Specialists from several nuclear power plants including Kalinin, Kursk and Smolensk also attended the workshop. The foreign delegation consisted of representatives from international agencies, national laboratories, industry, and academia. Each of the foreign participants presented papers and were key contributors to the workshop. This delegation consisted of Peter Chan, Atomic Energy of Canada, Ltd., John Magill of the EURATOM Transuranium Institute, Cyril de Turenne of Cogema, Erich Merz of the Juelich Research Center and the University of Aachen, Dieter von Ehrenstein and Roland Reimers of the University of Bremen, Vladimir Kagramanian of the IAEA, Lynn Farrington and Peter Broome of BNFL, Bruce Bevard of Oak Ridge National Laboratory and David Boyle, Ted Parish and Lee Peddicord of Texas A&M University.

The workshop was held at the Center for Education and Information Techniques of the Russian Employment Service (TSEZAN). TSEZAN is a wholly contained facility with guest rooms, cantine, meeting rooms, an auditorium and a lobby/social center. It is located on a wooded campus in the town of Ivanteevka, 35 kilometers northeast of Moscow. The meeting auditorium was equiped with public address and earphone systems which allowed for all of the papers presented in Russian to be simultaneously translated into English. The papers presented in English were interpreted into Russian sequentially.

The breadth and quality of the technical presentations, and the spirit of good will, friendship and collaboration which characterized the workshop, contributed immeasurably to the meeting's success. The key role of NATO was noted in both the opening and closing sessions.

Theodore A. Parish, Texas A&M University
Vyacheslav V. Khromov, Moscow Engineering Physics Institute
Igor Carron, Texas A&M University

ACKNOWLEDGMENTS

The organization of the workshop was the dual responsibility of the Moscow Engineering Physics Institute (MEPhI) and Texas A&M University (TAMU). MEPhI took care of most of the arrangements for organizing the meeting while TAMU primarily handled selecting the international delegation and editing the proceedings.

At MEPhI, a number of people were key to making the workshop a success. Professor Khromov served as the workshop Co-Director and provided top level guidance in the organization of the meeting. Due to his illness, a number of other people stepped forward to assume important roles. Professor Vladimir Naumov is acknowledged for compiling and editing the Russian language version of the meeting transactions. Special commendation also goes to Dr. Edward Kryuchkov, Dr. Vladimir Savander, Dr. Anatoli Chmelev, Mr. Pavel Tsvetkov, Mr. Vyacheslav Okunev, Mr. Yuri Mitjaev and Mr. Igor Zonov for their contributions in conducting the conference. Dr. Alexander Chebeskov of IPPE Obninsk is thanked for providing key assistance through his service on the steering committee. Professor Boris Onykii who had recently been elected as the new Rector of MEPhI is thanked for presenting the conference opening address. Of particular note was the quality of the interpreters, Ms. Ludmila Belatis and Mr. Sergei Yelovsky, who both possessed excellent familiarity with scientific terminology. Finally, the faculty, staff and students of the Department of Theoretical and Experimental Reactor Physics at MEPhI are thanked for all their efforts in tending to the requests/needs of the international delegation, and for generally making the workshop a success.

The support of the Nuclear Engineering Department at Texas A&M University made possible by its head, Dr. John W. Poston, Sr., is greatfully acknowleged. A number of individuals from the Nuclear Engineering Department were crucial to the workshop's success and deserve special mention here. The responsibilities for the workshop were primarily handled by Professor Parish, the workshop Co-Director, Dr. Igor Carron, and Ms. Gia Alexander. Professor Parish supervised the NATO grant and served as the chief editor of the proceedings. Dr. Igor Carron was instrumental in assuring the success of the workshop by taking care of many administrative details concerning travel reimbursement and financial reports. Dr. Carron also edited papers. Mr Bradley Rearden, a TAMU student, is acknowledged for the help he gave to Dr. Carron. Finally, Ms. Gia Alexander deserves special mention for her efforts which contributed to the success of the workshop. Her experience in editing earlier NATO proceedings was priceless in terms of advising authors, collecting the permission to publish forms and formatting the papers. Ms. Alexander was assisted by two TAMU students, Mr. Giby Joseph and Mr. Max Oyola.

Special thanks go to the MEPhI organizers, in particular, Dr. Edward Kryuchkov, Dr. Vladimir Savander and Mr. Pavel Tsvetkov, for hosting a number of social activities to provide the workshop participants with entertainment after the technical sessions. On Wednesday evening, there was a barbecue along with the singing of Russian folk songs. On Thursday afternoon, a trip was made by everyone to the Trinity Monastery in nearby Sergiev Posad. On Friday evening, there was a lively and festive banquet. And on Saturday afternoon, there was a visit to the recently reconstructed Christ the Savior Cathedral and a tour of Moscow at the height of its 850th birthday celebration.

This NATO advanced research workshop brought together Russian and western specialists to discuss questions associated with the safe elimination of both weapons-grade and reactor-grade plutonium in reactors. This workshop was the fourth in a series of workshops centered on plutonium disposition sponsored by the Scientific Affairs Division of NATO. Mrs. Nancy T. Schulte, Program Director of the Disarmament Technologies Division at NATO, gave invaluable and timely assistance.

The contributions of each participant in the meeting whether or not their papers are included in these proceedings is also thankfully acknowledged.

Theodore A. Parish, Texas A&M University
Vyacheslav V. Khromov, Moscow Engineering Physics Institute
Igor Carron, Texas A&M University

THE U.S.-RUSSIAN JOINT STUDIES ON USING POWER REACTORS TO DISPOSITION SURPLUS WEAPONS PLUTONIUM AS SPENT FUEL

A. CHEBESKOV
A. KALASHNIKOV
State Scientific Center—Institute of *Physics and Power Engineering*
I Bondarenko Sq.
Obninsk, Kaluga Region
249020 Russian Federation

B. BEVARD
D. MOSES
Oak Ridge. National Laboratory
PO Box 2009
Oak Ridge, Tennessee 3783
United States

A. PAVLOVICHEV
State Scientific Center—Kurchatov institute
I Kurchatov Sq.
123182 Moscow
Russian Federation

Abstract

In 1996, the United States and the Russian Federation completed an initial joint study of the candidate options for the disposition of surplus weapons plutonium in both countries The options included long-term storage, immobilization of the plutonium in glass or ceramic for geologic disposal, and the conversion of weapons plutonium to spent fuel in power reactors. For the latter option, the United States is only considering the use of existing light-water reactors (LWRs) with no new reactor construction or the use of Canadian deuterium-uranium (CANDU) heavy-water reactors. While Russia advocates building new reactors, the cost is high, and the continuing joint study of the Russian options is considering only the use of existing VVER-1000 LWRs in Russia, and possibly Ukraine, and the existing BN-600 fast-neutron reactor at the Beloyarsk

1

T. A. Parish et al. (eds.),
Safety Issues Associated with Plutonium Involvement in the Nuclear Fuel Cycle, 1–13.
© *1999 Kluwer Academic Publishers. Printed in the Netherlands.*

2

Nuclear Power Plant in Russia. The use of Canadian CANDU reactors is also an option. Six of the seven existing VVER-1000 reactors in Russia and the eleven VVER-1000 reactors in Ukraine are all of recent vintage and can be converted to use partial MOX cores. These existing VVER-1000 reactors are capable of converting almost 300 kg of surplus weapons plutonium to spent fuel each year with minimum nuclear power plant modifications. Higher core loads may be achievable in future years. The BN-600 reactor, which currently uses enriched uranium fuel, is capable (with certain design modifications) of converting up to 1,300 Kg or 1.3 metric tons (MT) of surplus weapons plutonium to spent fuel each year. The steps needed to convert the BN-600 to a plutonium-burner core are:

1. Elimination of the depleted uranium breeding blankets and their replacement with a combination of a steel reflector and boronated shield;
2. Initial conversion to a hybrid enriched uranium-plutonium-fueled core sufficient to preserve a zero value for the sodium void reactivity effect; and
3. Ultimate conversion to the plutonium-burner core that requires several modifications to the fuel design and the reactor.

The step involving the hybrid core allows an early and timely start that takes advantage of the limited capacity for fabricating uranium-plutonium mixed oxide (MOX) fuel early in the disposition program. Finally, the design lifetime of the BN-600 must safely and reliably be extended by 10 years to at least 2020 so that a sufficient amount of plutonium (~20 MT) can be converted to spent fuel.

1. Introduction

Significant quantities of weapons-usable fissile materials [primarily plutonium and highly enriched uranium (HEU)] are becoming surplus to national defense needs in both the United States and Russia. These stocks of fissile materials pose significant dangers to national and international security. The dangers exist not only in the potential proliferation of nuclear weapons but also in the potential for environmental, safety, and health (ES&H) consequences if surplus fissile materials are not properly managed.

The first and second Strategic Arms Reductions Treaties (START I and START II) call for deep reductions in the strategic nuclear forces of both the United States and the former Soviet Union. In addition, in the aftermath of the Cold War, both the United States and Russia have initiated unilateral steps to increase the pace of strategic disarmament. Under START and subsequent unilateral initiatives, some 10,000 to 20,000 warheads in the United States and a similar or greater number in the former Soviet Union) could possibly be declared "surplus" to national security needs. Thus, significant quantities of weapons-usable fissile materials have or will become surplus to national defense needs in both the United States and Russia.

On January 14, 1994, US President Clinton and Russian President Yeltsin issued a statement on Non-Proliferation of Weapons of Mass Destruction and The Means of Their Delivery, in which the Presidents tasked their experts to jointly "study options for the long-term disposition of fissile materials, particularly of plutonium, taking into account the issues of nonproliferation. environmental protection, safety, and technical and economic factors[1].

In 1996, the United States and the Russian Federation completed a joint study of the options for the disposition of surplus weapons plutonium in both countries [2]. The options included long-term storage, immobilization of the plutonium in glass or ceramic for geologic disposal, and the conversion of weapons plutonium to spent fuel in power reactors. For the latter option, the United States is only considering the use of existing LWRs with no new reactor construction for plutonium disposition. The Russian government's approach emphasizes use of plutonium as fuel for nuclear reactors because of its energy value. While Russia advocates building new reactors, the cost is high; an estimated $1.4 billion is needed to construct a BN-800 fast reactor. Therefore, the continuing joint study of the Russian options is considering only the use of the existing VVER-1000 LWRs in Russia and Ukraine, the use of Canadian deuterium-uranium (CANDU) heavy-water reactors in Canada, and the existing BN-600 fast neutron reactor at the Beloyarsk Nuclear Power Plant in Russia. This paper focuses on the use of the VVER-1000 and BN-600 reactors for disposition of surplus weapons plutonium in Russia.

2. The VVER Reactors

The isotopic composition of weapons-grade (or weapons-derived) mixed-oxide (MOX) fuel differs inherently from that of commercial reactor-grade MOX because weapons-grade plutonium has higher fissile content and lower ^{240}Pu content than reactor-grade plutonium. This difference is not expected to affect either the VVER- 1000 fuel assembly configuration or the reactor performance of the MOX fuel. The reference conversion process for MOX fuel production from weapons-grade plutonium in the Russian Federation is expected to be aqueous conversion with purification such that the powder is chemically identical to that used commercially in reactor-grade MOX. Use of MOX fuel made from reactor-grade plutonium in LWRs is already under way in Europe on a substantial scale, with 34 reactors now licensed for MOX fuel use [3], and is planned to begin soon in Japan. Therefore, the technical feasibility of using MOX in LWRs is amply demonstrated. Although neither the United States nor Russia has any LWRs currently using such fuel, both have LWRs already in operation that may be suitable for using plutonium in the form of uranium-plutonium MOX fuel. The use of such reactors would allow weapons plutonium to be transformed into spent fuel in a timely fashion. This could begin within 5 years from a decision to undertake such a project and would extend over a period of 10 to 20 years thereafter. The fissile material in spent MOX fuel would be roughly as difficult to recover for use in nuclear weapons as the fissile material remaining in low-enriched uranium (LEU) spent fuel.

The use of MOX fuel changes the physics of the reactor core significantly compared to the uranium fuel usually employed, and it is essential to ensure that nuclear safety is maintained if MOX fuel is to be used. Traditionally, most LWRs that have used MOX fuel have used it in only one-third of their fuel assemblies to limit the change in safety parameters compared to using uranium fuels. Using MOX in larger fractions, up to 100% of the core, is possible if adequate attention is paid to ensuring effective control of the reactor. Full MOX cores would have the advantage of greatly reducing the number of reactors needed to accomplish disposition of a given amount of plutonium in a certain period of time and therefore reducing the necessary transportation of fuel containing weapons-grade plutonium and the number of sites handling such fuel. Belgium has demonstrated the use of a 70% MOX in an experimental reactor; three operating U.S. reactors were specially designed for 100% MOX cores although they have not been demonstrated or licensed in this mode; and a substantial number of other U.S. reactors are believed capable of full MOX core operation.

The United States has some past experience with LWR MOX dating from the 1950s, well before the 1976 U.S. decision not to pursue near-term plutonium separation and recycle. Computer codes for modeling the behavior of LWR reactor cores with MOX fuel are available and are being compared to existing Russian codes. Initial fuel development tests, in which MOX fuel rods containing weapons plutonium will be irradiated in test reactors simulating the conditions in a commercial VVER, are scheduled to begin in 1997-1998. Information gained in these tests will be used to help validate these computer codes.

Russia has no experience with the use of MOX in its LWRs because its plutonium fuel plans have been traditionally focused on fast-neutron reactors. The use of MOX in LWRs is now being studied however, and Russia may be able to make use of MOX experience in Europe. There are seven operational VVER-1000 reactors in Russia of which six are considered capable of supporting the plutonium disposition mission. Two more VVER-1000s are under active construction and are expected to be completed in the near future: they are estimated by Russia to be 80-90% complete. A third new VVER-1000 reactor, estimated to be 70% complete, has less current construction activity under way and is expected to be completed by 2003 if adequate financing is available. Two additional VVER- 1000s and a number of the new VVER-640s are planned, but the availability of financing for these projects is uncertain. In addition to the VVER-1000 reactors in Russia, there are potentially 11 Russian-designed VVER-1000 reactors in Ukraine that may be available for the plutonium disposition program. These reactors were constructed from the 1980s through the mid-1990s and are believed to meet most Western safety standards. Thus, significant reactor modifications are not expected to be needed to convert from LEU fuel to partial MOX fuel.

In both the United States and Russia, the major factors determining when this option could begin are the need to provide the necessary fuel fabrication facilities and the need to acquire licenses and political approvals for both those facilities and the reactors that would use plutonium. To the extent possible, all alternatives would make use of existing infrastructure and capabilities at Russian nuclear sites. This approach would minimize cost and provide new missions for existing facilities, manpower, and intellectual resources rendered idle by the end of plutonium production for weapons.

Preliminary studies are under way on VVER-1000 reactors with one-third MOX cores to determine the extent of reactor modifications that may be necessary. Plutonium used as a fuel results in a more negative cooling water temperature reactivity coefficient and reduced boron efficiency. Control rod efficiency, boric acid concentration, and the rate of boric acid injection into the primary circuit under emergency conditions become the most important parameters to determine how many subassemblies are allowed to have MOX fuel. Modifications to the reactor safety systems could include increasing the diameter of the control rods, changing the material from which they are made, or adding more control rods. Preliminary designs allow for an increase in the number of control rods from 61 to 121 (the reactor design permits this upgrading) and introduction of new monitoring and diagnostic systems. To increase control rod efficiency in VVER-1000s, modifications could also include increasing the number of absorber rods in an assembly from 18 to 24 and increasing the boron enrichment of the ^{10}B isotope in the absorber rods. It may be easiest to increase the absorber diameter. Preliminary investigation shows that it is possible to increase the absorber diameter from 7.0 to 7.6 mm with a simultaneous increase in the guide tube outer diameter from 12.6 to 13.1 mm. This improves the rod system efficiency by ~6% [2]. Another safety improvement option, not requiring reactor redesign, is to use a core reloading scheme with lower neutron leakage. In this scheme, part of the fuel assemblies with fresh fuel are loaded into the central part of the core. It is important to use fuel rods with gadolinium burnable poison. Along with flattening of the core power distribution, this loading scheme allows the neutron flux to rise in the fuel assemblies with control rods and hence to increase the negative reactivity worth of the rods near the end of the reactor cycle, when it is most needed. Whether it is possible to increase the percentage of the core loaded with MOX fuel to 50, 75, or 100% without substantial and costly modifications to the reactor requires further study.

The planned new-design reactors (VVER-640s) should be able to handle full MOX cores safely because they will employ twice the number of control rods used in most existing VVER-1000s. The following passive safety systems are also planned to be installed in the new VVER-640 reactors:

- Core heat removal for use during reliable power supply failure (PCHRS);
- Core flooding for accidents with blackout and primary circuit leaks;
- Catching, confining, and cooling corium after reactor vessel melt-through;
- Gas-vapor filtration for emergency discharge into the environment during an unanticipated pressure rise of more than 5 atm inside the containment; and
- Double containment (steel and concrete).

Additionally, the following measures may be taken to reduce exposure for plant maintenance personnel when converting VVER-1000s to MOX fuel:

1. Construct separate storage for fresh MOX fuel at the nuclear power plant, designed for the MOX fuel for all reactors. This storage must have a MOX fuel subassembly inspection bay and facilities for loading the subassemblies into on-site containers;
2. Develop on-site containers; and
3. Develop fresh MOX fuel containers and transportation equipment.

Spent MOX fuel subassemblies submersed in water have a higher neutron multiplication factor than spent uranium fuel subassemblies. Therefore, it is necessary to increase the lattice pitch of the spent fuel storage pond rack, or the rack needs to be made of structural steel containing boron or other elements with high neutron-absorbing properties. The spent MOX fuel container and the methods for transporting and storing spent MOX fuel are similar to those for spent uranium fuel. However, more long-term cooling of the spent MOX fuel assemblies is required at the nuclear power plant before the assemblies can be shipped to permanent storage facilities.

Russia has pilot-scale MOX fabrication facilities at Mayak and Dmitrovgrad, which are capable, after some redesign, of producing small amounts of LWR MOX fuel for experimental purposes. Russia is currently collaborating with European partners on the conceptual design of an expanded pilot plant at Mayak with a capacity of 1.3 MT of plutonium per year. This is enough to provide partial MOX cores for four VVER- 1000 reactors and for the BN-600 fast-neutron reactor. Several options for commercial-scale production of LWR MOX exist. Current Russian plans, subject to the availability of financing, call for construction of a MOX plant dedicated to producing LWR fuel beginning after the turn of the century, in conjunction with the planned RT-2 reprocessing plant at Krasnoyarsk-26. Alternatively, the partially completed "Complex-300" MOX plant at Mayak could be finished and one of the lines modified for production of LWR MOX, or a new facility could be built at that site. Further study of the costs, schedules, and nonproliferation and safety implications of each of these approaches is needed.

Assessing total program costs of the LWR option in Russia is very difficult because Russia's rapidly changing economic circumstances introduce substantial uncertainties into any long-term economic assessment. It is apparent that the small amount of NPP modifications and infrastructure changes necessary to use existing VVER-1000s would cost significantly less than building new NPPs. Current estimates reflect a cost for using these NPPs at a level similar to the cost of immobilizing the weapons-grade plutonium, but with the added advantage of realizing the electrical power potential of the plutonium. Russia is currently considering a substantial MOX program designed to manage the civilian plutonium arising from reprocessing. Financing of this program is uncertain. Therefore, the cost assigned to disposition of weapons plutonium by the MOX route should be the net additional cost of modifying the previously envisioned MOX program to handle both weapons plutonium and civilian plutonium. However, it is also important to identify the needed capital investments for any MOX program. This will facilitate planning for the necessary financing for disposition of either civilian or surplus weapons plutonium.

3. The BN-600 Reactor

Currently, BN-600 is fueled with enriched uranium and is a demonstration "breeder" reactor although its current operations, which are directed at producing electrical energy, are not optimized to make it an efficient producer of fissile plutonium compared to the consumption of fissile uranium fuel. However, the ~100 blanket assemblies removed each year contain ~120 kg of plutonium with about 95% ^{239}Pu. The BN-600 reactor is capable, with certain design modifications, of being converted from a plutonium producer to a net burner of plutonium that can disposition up to 1.3 MT of weapons plutonium into highly radioactive spent fuel each year.

The BN-600 reactor is currently licensed by the Russian Federal Nuclear and Radiation Safety Authority (GOSATOMNADZOR or GAN) to operate with 18 fuel subassemblies containing MOX fuel elements in a core of 369 subassemblies that are normally fueled with enriched uranium oxide. To date, 24 MOX fuel subassemblies have been irradiated in the BN-600. Of these subassemblies, 6 contained vibro-packed MOX fuel fabricated at the Research Institute of Atomic Reactors (RIAR) in Dmitrovgrad, and the other 18 used pelletized MOX fuel fabricated at the PAKET pilot plant at Mayak, Chelyabinsk Region. The fuel in the BN-600 tests used plutonium oxide from reprocessed radial blanket subassemblies from BN-350 and BN-600 so that the plutonium isotopic composition is very close to that of weapons-derived plutonium. The irradiations in BN-600 supplement the extensive prior testing of plutonium oxide and MOX fuels at the BR-10, BOA-60, and BN-350 fast-neutron reactors.

The steps needed to convert BN-600 to a full MOX, plutonium-burner core are as follows: (1) elimination of the radial breeding blanket and its replacement with a combination of a steel reflector and boronated shield, (2) initial conversion to a hybrid core (based on a predominantly uranium-fueled core partly loaded with MOX fuel) sufficient to preserve a zero value for the sodium void reactivity effect (SVRE), and (3) ultimate conversion to the full MOX core. The hybrid core conversion requires a fuel fabrication facility capable of supplying MOX fuel using ~300 kg/year of surplus weapons-derived plutonium. The full MOX core requires modifications to the design of the fuel subassembly to obtain a negative SVRE value, reduction of the sodium pump head by modifying the main coolant pumps to accommodate the modified fuel subassemblies, and a MOX fuel fabrication capacity using ~1.3 MT/year of surplus weapons-derived plutonium and dedicated to BN-600.

The BN-600 reactor will reach the end of its initially planned design lifetime in 2010. To make a significant contribution to plutonium disposition (~20 MT), the lifetime of the BN-600 must safely and reliably be extended to at least 2020. The BN-600 power plant has an aggressive in-service inspection program to monitor plant aging effects in structures and components. Life extension is judged to be feasible because the plant is in excellent condition and suppliers of replacement equipment exist. The BN-600 power plant judges the limits to extended life to be tied to the financial situation in Russia. not to any technical or safety-related restrictions.

The first step in reconfiguring the BN-600 to become a plutonium burner is to eliminate the radial breeding blanket that surrounds the core and separates the core from the in-vessel spent fuel storage. The radial blanket consists of ~400 subassemblies fueled with steel clad rods containing depleted uranium oxide pellets. About 100 of the subassemblies in the radial breeding blankets are removed each year. These contain ~120 kg of plutonium with about 95% ^{239}Pu. However, the blanket is also needed to attenuate the neutrons leaking from the core into the in-vessel spent fuel storage area so that fission heating in the stored fuel is acceptably low. In recent years, the Russian RT-1 reprocessing plant at Chelyabinsk has ceased to accept the radial blanket subassemblies for reprocessing. Currently, there is sufficient space in the BN-600 water-cooled ex-vessel spent fuel storage pool for about 3 years, then alternative storage for the irradiated blanket subassemblies will have to be found, or the reactor may have to be shut down.

The optimum solution involves the elimination of the radial breeding blankets and the construction of a dry storage facility for previously irradiated blanket subassemblies. The current inventory in wet storage contains an estimated one metric ton of weapons-quality plutonium. The irradiated blanket subassemblies are substantially less radioactive than the irradiated fuel subassemblies. Axial breeding blankets are integral with the fuel rods in the fuel subassemblies that are highly radioactive after irradiation.

To eliminate the radial breeding blanket, several design changes are required for the core. Steel reflector subassemblies must be designed and fabricated to replace the radial breeding blanket subassemblies immediately surrounding the core. Similar subassemblies are used as gamma shielding in the BN-600 around the base of the refueling elevator outside the radial blanket, but the conceptual design would use different locations adjacent to the core. The candidate material is 12% chromium, 1% molybdenum ferric stainless steel, which has a lifetime neutron fluence limit of 120 displacements per atom based on testing at Dmitrovgrad. Such subassemblies have also been used in the United States both at the Experimental Breeder Reactor II in Idaho and at the Fast Flux Test Facility at Hanford, Washington. Shield subassemblies must be designed and fabricated to replace the radial breeding blanket subassemblies in the outer locations adjacent to the in-vessel spent fuel storage. Neutron leakage radially from the core to the spent fuel must be attenuated by the shield subassemblies in a manner comparable to the radial breeding blanket so that an acceptably low level of subcritical fission heating is maintained in the in-vessel stored spent fuel. The conceptual design of the shield subassemblies is for steel-clad rods containing boron carbide pellets to moderate and capture neutrons leaking past the reflector subassemblies.

The core must be enlarged slightly by adding ~20 fuel subassemblies to compensate for power generation lost by removing ~400 subassemblies from the radial breeding blanket. Compared to fuel subassemblies, radial blanket subassemblies have a different inlet orificing in the extension on the lower part of the subassembly to reduce flow. Adding 20 fuel subassemblies with higher flow and ~380 reflector/shield subassemblies with slightly reduced flows is calculated by the designers to not pose a problem from the standpoint of the thermal-hydraulic margin of safety.

The elimination of the radial breeding blanket can proceed prior to or in parallel with the conversion to the hybrid partial MOX core. The important issues are to eliminate the production of ~120 kg of weapons-capable plutonium (as judged from its isotopic composition) each year in the blanket, to ensure that the margin of safety in the reactor is not compromised, and to secure in safe storage the ~1 MT of weapons-capable plutonium contained in irradiated radial breeding blanket subassemblies. The plan is to solve the problems of breeding blanket elimination and storage before 2001.

3.1. CONVERSION TO A HYBRID (PARTIAL MOX) CORE

The BN-600 core is licensed by GAN to contain up to 18 MOX subassemblies at any one time. The principal regulatory limit to adding additional MOX subassemblies without significantly changing the current fuel subassembly design is related to maintaining a nonpositive value for SVRE. Because of the reactivity transient that occurred in the Chernobyl accident, the GAN regulations prohibit positive reactivity feedback due to voiding of the coolant. In the uranium cores of BN-600 even with a few MOX subassemblies, the SVRE value is strongly negative. As additional MOX subassemblies are added to the core, calculations show that the SVRE value becomes less negative and, at around 90 subassemblies or so (depending on the zoning arrangement), the SVRE value is close to zero. To meet the GAN requirements and to ensure that SVRE is at most zero, or a negligibly small positive value, the designer must select a design that provides a sufficiently negative calculated value of SVRE to compensate for the uncertainties in calculations and experimental benchmarks. While applying the deterministic SVRE criteria in the design of the hybrid core, this effort will be supplemented by probabilistic safety analyses to demonstrate that the probabilities and consequences of total or partial core voiding are acceptably small for the hybrid core. The GAN licensing is expected to take about 3 years with simultaneous review of the safety case for elimination of the radial breeding blanket. The current planning is to initiate BN-600 operations with a hybrid MOX core by 2002.

In addition to the design and safety studies on the behavior of the hybrid core during normal operations and accidents, which will be documented in the updated safety analysis report submitted to GAN, an adequate capacity for supplying reload MOX subassemblies must be developed and licensed. The initial hybrid core loading will require 70-90 MOX subassemblies, and core reloads will require 40-50 MOX subassemblies per year using ~300 kg of surplus weapons-derived plutonium annually. BN-600 has favorable irradiation experience with both vibro-packed and pelletized MOX subassemblies using reprocessed plutonium oxide from BN-350 and BN-600 radial breeding blankets and containing about 95% ^{239}Pu. Several options are being considered for interim MOX fabrication capacity to support the hybrid core.

3.1.1 Upgraded PAKET pilot line at Mayak in Chelyabinsk Region

This option would upgrade and expand the Russian facilities used currently to make the four subassembly batches of MOX fuel for BN-600. Currently, rod bundling of the four subassembly batches takes place at Elektrostal near Moscow. However, for 40-50 subassemblies per year, this capability would be replicated on a small scale either at the Mayak site or at RIAR where licensed plutonium-handling facilities already exist. Collocating all fabrication facilities at Mayak would minimize transportation of fissile materials between sites and place the fabrication facilities on the same site as the dismantled weapon storage facility. At PAKET, conversion of weapons-derived metal into an oxide powder would be based on aqueous processing such as either an oxalate precipitation of plutonium oxide with subsequent mechanical mixing with uranium oxide powder or ammonia coprecipitation of MOX powder. Small-scale facilities for each process already exist at Mayak, and both types of pelletized fuel have been irradiated in BN-600 with excellent performance.

3.1.2 Expanded vibro-packed capacity at RIAR in Dmitrovgrad

RIAR currently has facilities for the recycling of civilian plutonium from BOR-60 reactor fuel, but expanded facilities would be needed to provide the annual requirement for 40-50 subassemblies for the BN-600 hybrid core. RIAR uses pyroelectrochemical processing in a molten salt to produce the powder for vibro-packed fuel, which issued in BOR-60 and has been tested in BN-600. This technology can be applied to the conversion of weapons-derived metal or oxide into MOX. The disadvantage of collocating all fabrication facilities at RIAR is that weapons-derived metal or oxide from Mayak would have to be transported to RIAR. The production of oxide powder at Mayak would reduce the attractiveness of the material to theft or diversion during transport to RIAR from Mayak, but it also introduces an additional, unnecessary first step from the standpoint of fuel performance, requires additional accident analysis of potential contamination events in transit, and complicates material control and accountability.

3.1.3 TOMOX-DEMOX

From 1993-1996, the French and Russians worked on a joint project on plutonium disposition designated AIDA MOX Phase 1; AIDA MOX Phase 2 is now starting. The products of this effort include conceptual designs for a plutonium metal-to-oxide conversion pilot facility (TOMOX) and a MOX fuel fabrication pilot facility (DEMOX) with a capacity of 1.3 MT/year of plutonium metal. The vision for use of these facilities is to process 300 kg of surplus weapons plutonium into MOX fuel for BN-600 and 1000 kg for VVER-1000 fuel subassemblies. Thus, TOMOX-DEMOX would provide fuel for one fast reactor and about four water reactors. The full MOX option in BN-600 requires the dedication of a facility of equal capacity to TOMOX-DEMOX. The current reality of this proposal is that it has a split mission (BN-600 and VVER-1000), lacks consensus on location (Mayak or Krasnoyarsk), and lacks consensus on processes with France, the United States, and several Russian institutes advocating varying technologies, especially for TOMOX. Without arriving soon at a consensus favorable to the BN-600 mission, it is likely that this approach may not be sufficiently timely to support an early start of the hybrid core conversion. However, the upgraded PAKET option may also be subsumed by this proposal due to limits on Western financing of a pilot plant.

An additional concern, raised by personnel of the BN-600 power plant and core designers, with regard to using surplus weapons plutonium is the possible need for changes in the reactor fresh fuel handling and shielding systems to accommodate the higher gamma-ray source from ^{241}Am in the weapons plutonium. Specialists from the BN-600 power plant indicate that the measured radioactive exposure dose from experimental MOX subassemblies made from plutonium reprocessed from BN-350 and BN-600 radial blankets is higher than the exposure dose from conventional uranium subassemblies. Operational procedures at the BN-600 stipulate that appropriate measures should be taken to protect workers during handling operations with MOX fuel subassemblies.

3.2. CONVERSION TO A FULL MOX CORE

Conversion of BN-600 to a full MOX plutonium-burner core requires design changes in the reactor system to ensure an acceptable nonpositive or negligibly small positive SVRE value. In addition, adequate fuel supply capacity is needed to provide a sufficient number of MOX subassemblies containing ~1.3 MT of surplus weapons-derived plutonium each year. The intent is to complete conversion to full MOX between years 2005 and 2007 so that ~20 MT or more of surplus weapons plutonium can be consumed and transmuted to spent fuel before 2020.

As discussed previously, replacing more than about 90 of the enriched uranium fuel subassemblies with MOX fuel subassemblies leads to a positive SVRE value when using the current subassembly design for BN-600. This problem was solved analytically, in conjunction with experiments in the BFS fast reactor critical facility, for the next-generation Russian fast reactor (BN-800) by modifying the design of the rod bundle within the subassembly can. This approach, which can be adapted to BN-600, is based on eliminating the upper axial breeding blanket in the fuel rods, introducing a sodium plenum immediately above the fuel rod bundle, and placing a cluster of boronated, short rods above the plenum within the subassembly can. In this case, the introduction of the sodium plenum requires reducing the core height by about 150 mm less than in the hybrid core. With this design, voiding that initiates in the hottest flow channels of the upper core rises to the upper plenum, displacing liquid sodium that serves as a neutron reflector and producing increased neutron leakage from the top of the core into the boronated shield. The loss of neutrons in such a scenario creates a negative SVRE value.

Because of the loss of fission heating due to shortening of the core height and removal of the upper axial breeding blanket, the radial size of the BN-600 full MOX core will also have to increase by about 35 subassemblies compared to the hybrid core to maintain the same power generation capacity for the plant and the same thermal performance margin of safety in the core. In addition, the removal of the lower axial breeding blanket may be desirable from the standpoint of further improving BN-600 as a net burner of plutonium, but it is not considered practical at this time and would require substantial further study.

An adequate supply of MOX fuel is needed to continue the BN-600 on full operations as a plutonium burner until its end of life. As indicated, the French-Russian TOMOX-DEMOX project is for a pilot plant with the requisite capacity for BN-600 on full MOX, but it is currently envisioned to provide VVER-1000 fuel also. The capacity of the TOMOX-DEMOX pilot plant is not sufficient to supply the needs of the BN-600 on full MOX ,and as many as 7 VVER-1000s in Russia, and possibly 11 VVER-1000s in Ukraine, on partial MOX. However, at this time, it is not yet clear how many plutonium conversion facilities and MOX fabrication facilities will be constructed and at what capacity. This issue is currently being addressed separately in bilateral discussions between Russia and France, Germany, and the United States, respectively. Thus, a major uncertainty for the BN-600 full MOX option is the timing of the fuel supply.

3.3. EXTENSION OF THE OPERATING LIFETIME FOR BN-600

As indicated previously, BN-600 has a predicted design lifetime of 30 years ending in 2010. The predictions are based on conservative estimates of materials and structural performance in nonreplaceable components and high-cost components. The BN-600 power plant has an aggressive in-service inspection and maintenance program and has replaced steam generator evaporator modules, which performed as predicted. An intermediate loop heat exchanger will be removed and inspected for evidence of age-related degradation phenomena before 2000. The steam generator superheaters in the intermediate loop must also be inspected prior to the end of their conservatively predicted design life in 2010. Suppliers exist for all key components. Financing replacement equipment procurement is the only issue for the BN-600 power plant.

The Experimental Design Bureau of Mechanical Engineering (OKBM) maintains an operational data base on thermal-hydraulics and structural-mechanical performance. OKBM has used measured neutron fluence data to validate its lifetime predictions for neutron-irradiated reactor components. The lifetime margins of irradiated structures vary from a factor of 1.6 for the rails of the in-vessel refueling elevator, which are replaceable, to a factor of 8 to 20 for the reactor vessel. The core barrel, which is not load bearing, is highly irradiated but not life limiting. The reactor coolant pump impellers have been redesigned for extended life. The major impediment to life extension is the availability of financing in a timely manner to support the procurement of needed replacement equipment.

4. Conclusions

The use of existing VVER- 1000 reactors and the BN-600 fast reactor for the disposition of surplus weapons-grade plutonium into spent nuclear fuel is a technically viable option. Compared to the construction of new reactors, the use of modified VVER-1000s and the BN-600 reactor (with an extended lifetime) offers a less expensive and more timely alternative for disposition that takes full advantage of existing facilities and equipment. Compared to the immobilization alternative, the existing reactor option has the significant advantage of using the enormous energy potential of plutonium. The technical and regulatory problems to be solved are tractable. The United States and the Russian Federation have currently embarked upon the planning and preliminary analyses needed to execute the work necessary to use these reactors in a timely and safe manner.

5. Acknowledgments

Research sponsored jointly by the Amarillo National Resource Center for Plutonium and the Office of Fissile Materials Disposition, U S Department of Energy, under contract DE-AC05-960R22464 with Lockheed Martin Energy Research Corp.

References

1. *Non-Proliferation of Weapons of Mass Destruction and the Means of Their Delivery,* Joint Statement by U.S. President Clinton and Russian President Yeltsin, January 14, 1994 (cited in Ref. 2).
2. Office of Science and Technology Policy, U.S. Department of Energy; and the Russian Federation Ministry of Atomic Energy, *Joint United States/Russian Plutonium Disposition Study,* Government Printing Office, September 1996.
3. G. Le Bastard, "L'utilisation du MOX dans le monde," presented at the SEEN Conference on 1987-1997: 10 ans de Combustibles MOX en France, Paris, June 17, 1997.

CURRENT STATE OF AND NEAR TERM PROSPECTS FOR PLUTONIUM MANAGEMENT IN RUSSIA

A.M. DMITRIEV
Federal Nuclear and Radiation Safety Authority of Russia
RF GOSATOMNADZOR
Russia 109147, Moscow, Taganskaya ul., 34
Tel.:+095-911-64-13, 278-04-86,   Fax:+095-912-12-23

Abstract

Issues related to safe plutonium management and to the prospects for using the plutonium extracted from weapons as reactor fuel are becoming more and more important for the industrial development, environmental protection and national security policies of Russia. In recent years, because of the continuing decrease in the threat of nuclear confrontation, significant amounts of nuclear materials that have been extracted and accumulated at nuclear weapons related facilities as a result of dismantling warheads. These materials consist of both highly enriched uranium and weapons-grade plutonium.

1. Introduction

The question of what to do with highly enriched uranium from dismantled weapons finds its answer in chemical processing and blending with natural or depleted uranium to obtain uranium-hexafluoride with ^{235}U enrichments in the range of 4 to 5 %. Such uranium-hexafluoride is suitable for utilization with the standard technology used to prepare fuel for commercial nuclear powerplants. Simultaneously, some of the ex-weapons uranium is to be exported to the United States under a long-term agreement and a considerable part will be stored for further possible utilization in two ways. These are 1) sale of uranium-hexafluoride abroad and 2) storage for eventual use as nuclear fuel in Russian nuclear powerplants after 2010 to 2020. The technology for the preparation of uranium-hexafluoride with enrichments of 4 to 5 % in ^{235}U from highly-enriched weapons uranium has already been accomplished on a practical scale with the preparation of reactor fuel by Russian industry.

15

T. A. Parish et al. (eds.),
Safety Issues Associated with Plutonium Involvement in the Nuclear Fuel Cycle, 15–23.
© 1999 *Kluwer Academic Publishers. Printed in the Netherlands.*

16

The situation with respect to plutonium management/disposition is a bit more uncertain. Twenty to thirty years ago, the Russian scientific institutes of nuclear technology were anxious regarding the existence of proven, large, domestic uranium resources. This circumstance led to intensive research into nuclear technology with the goal of nuclear fuel breeding through the development of fast breeder reactors using sodium as a coolant. However, the actual scale of plutonium utilization in such reactors to date has been relatively insignificant.

Up until the end of 1997, Russia has been obliged to announce the amount of weapons-grade plutonium which is to be counted as surplus from a national security point of view. Preliminary estimates indicate that this amount will be about 50 tons. In addition to excess weapons-grade plutonium, Russia has some 30 tons of separated civilian plutonium at the PT-1 (the Mayak complex) reprocessing plant. This material was stored after separation from spent nuclear fuel from Russian VVER-440 nuclear powerplants. In declaring the amount of surplus plutonium, it has to also be taken into account that Russian double-purpose uranium-graphite reactors continue to produce (unseparated) weapons-grade plutonium in Seversk and Jeleznogorsk. Within the framework of the US - Russia joint agreement, this plutonium is not used for the Russian military complex as of October 10, 1994. As a result, this plutonium is added every year to the plutonium inventory which is declared to be in excess for the purposes of Russian national security. The plutonium production capabilities of the cores of the above mentioned reactors are being studied in order that the operation of these reactors as commercial sources of heat and electricity can be continued while at the same time reducing the amount of plutonium produced. After the double-purpose uranium-graphite reactors have been modified, the reduction of the plutonium production is expected to be about 20 - 30 times and the plutonium produced in the modified reactors will not meet the standard for weapons-grade. However, before the year 2000, it is impossible to perform the conversion of the uranium-graphite reactors, and hence, the above mentioned reactors will add several tons of weapons-grade plutonium to the already existing amount.

Thus, the problems related to plutonium management in general, and in particular, to weapons-grade plutonium are rather complex. The management of weapons-grade plutonium has the highest priority for two reasons. The first is that its nuclear characteristics make its usage simpler within the weapons fabrication process, and hence, weapons-grade plutonium is more attractive, as compared to reactor-grade plutonium, to those people and/or organizations who are interested in the construction of nuclear weapons. The second reason is that weapons-grade plutonium, as a rule, is in a metallic form that is more unstable and therefore, requires additional measures for safe storage.

In January 1995, a joint meeting of expert groups from the US and Russia was convened at Los Alamos to determine priorities for plutonium management from the following possibilities:
- burning in fast breeder reactors;
- burning in pressurized water reactors;
- burning in CANDU reactors;
- storage in special repositories;

- transmutation in electro-nuclear (accelerator-driven neutron source) units;
- immobilization in systems using special glass or minerals;
- disposal in geologic repositories; and
- techniques for handling plutonium in complex chemical/physical forms.

After discussions and detailed investigations of all of these possibilities for plutonium management, only two priority areas were selected for further study:

- burning in reactors (all types); and
- storage.

2. Plutonium Disposition Policy in Russia

The question of how to accomplish management in the case of plutonium in complex chemical/physical forms remains open, and will probably remain open for the foreseeable future as well. This problem was removed from the official list of priority areas because complex forms of plutonium cannot be used as market products and they require processing that is too complicated for their easy reuse in weapons, and therefore in effect, such plutonium is already to an appreciable degree immobilized.

From the very beginning, the Russian side has considered plutonium burning in various types of reactors as the most attractive method for its disposition. In Russia, this point of view is common for both organizations which utilize atomic energy and organizations which work in the field of nuclear safety regulation. Even with the selection of the same method (bruning in reactors) for elimination of excess plutonium, the instinctive motivations of the Ministry of the Russian Federation for Atomic Energy (Minatom) and those of the GOSATOMNADZOR are in principle different.

Minatom's representatives declare that plutonium is an important energy resource, and hence, plutonium burning is necessary to realize its economic potential, and that this potential energy resource is needed to support the national economy. However, this point of view can be easily criticized. In actuality, the fissioning of one gram of plutonium provides a thermal energy release which is equivalent to the burning of 2.8 tons of coal. Thus, one hundred tons of plutonium is equivalent to 2800000 tons of good quality coal or an amount approximately equal to one half of one year's coal mining in the USSR at the level in 1980. As a result, the complete burning of the existing amount of excess plutonium cannot solve the energy problems of Russia. At the same time, development of the specialized industry, which will be needed to perform processing, fabrication, transportation, and burning of plutonium fuel, and further chemical reprocessing of the irradiated fuel within a closed fuel cycle, will require very high levels of near-term capital investment under any scenario with a reasonable timetable for burning.

The GOSATOMNADZOR supports burning of plutonium as a way of managing plutonium because at least currently in Russia immobilization of plutonium in systems using special glass cannot satisfactorily guarantee sustainable long-term disposition. Implementation of such a method for disposition may create problems if the glass breaks exposing the plutonium inside. It needs to also be taken into account that immobilization, in principle, leaves open the theoretical possibility of subsequently extracting plutonium from the immobilized material. For immobilization to be viable, realistic extraction processes should be complicated and expensive to accomplish. However, even the existence of possible plutonium extraction processes, is not desirable from a non-proliferation point of view.

In the case of plutonium burial in deep boreholes, the geo-chemical processes are very complicated in the bowels of the earth, and hence, it is difficult to guarantee the total impossibility of plutonium migration and subsequent release into the environment over tens of thousands of years. It is much more simple to deal with fission products because significant experience has already been accumulated in the field of fission product management and the life-times of the main nuclides, which account for most of the radiological hazard, are much less than the life-time of plutonium.

In Russia, the views on plutonium management are also dependent on the financial resources which will have to be expended to achieve a sufficient level of safety during all of the operations for plutonium use/storage.

3. Temporary Storage

In Russia, storage of most of the plutonium inventory is a completely unavoidable step in the secure management of the accumulated stockpiles of both weapons-grade and reactor-grade plutonium. Storage of weapons-grade plutonium in metallic form is considerably more expensive than storage of plutonium in oxide form. However, long-term storage of plutonium is planned for the (di)oxide form as well as for the metallic form because, currently in Russia, there is no capability to perform timely and inexpensive conversion of metallic plutonium to the oxide form. Also, prior to storage of metallic plutonium which is obtained from weapons components, it is necessary to change the shape and size of the pieces in order to destroy any information regarding the design of the dismantled weapons.

Minimum estimates for the cost of storage of metallic plutonium in special repositories are at the level of US$ 0.5 per year per gram of plutonium. Currently, a prototype repository for plutonium and highly-enriched uranium is being constructed at the Mayak complex. The repository is a complicated engineering facility which must satisfy stringent requirements for physical security, monitoring and maintenance. The lifetime of the repository is projected to be one hundred years, and prior to its end of life, the nuclear materials will be removed for further processing. In addition, more than twenty thousand special containers have been fabricated for storage of special nuclear materials. This work has been performed with considerable technological and financial support from the United States.

4. Reactor Capabilities for Plutonium Utilization

Practical experience with MOX fuel utilization has been accrued in Russia only in fast, sodium-cooled reactors. If small-scale research installations are not taken into account, there is only one commercial fast reactor that is currently operable in Russia, the BN-600 reactor at the Beloyarskaya nuclear powerplant. Currently, MOX fuel constitutes 7 % of the BN-600 reactor core and it is loaded on the periphery. MOX fuel has never been loaded into the highly-enriched central fuel region of the BN-600 reactor core.

MOX fuel has not yet reached, on average, its projected level of burnup. The high levels the capital and operating expenses of fast reactors has meant that the financial support to build two BN-800 reactors, which had been expected, is currently not available.

Because there is no reason to presume that BN-800 reactors will be built for at least 10 to 12 years, the possibility of conversion of the the currently operating BN-600 reactor from a breeder core to a fully loaded MOX core for plutonium burning has been studied during recent years. The conversion of the BN-600 reactor to use a full core of MOX fuel could provide a timely way to reduce the amount of excess weapons-grade plutonium in Russia.

The existing capacity for MOX fuel production is not sufficient to provide the annual needs of the BN-600 reactor in the case of its full conversion from uranium oxide to MOX fuel. Therefore negotiations are being undertaken on a project to construct a new facility for MOX fuel production which will be able to produce MOX fuel for fast reactors, as well as for VVER reactors.

In the meantime, it is planned to first produce three fuel assemblies for VVER-1000 reactors which will be tested at the Balakovskaya nuclear powerplant. An agreement on production of the first MOX fuel pellets for CANDU reactors has already been signed. Initial estimates for the applicability of this approach for reducing the Russian amount of excess weapons-grade plutonium can be expected in 1998. In any case, only pressurized water reactors using MOX fuel can provide a high rate of consumption for excess plutonium during the next 5 to 10 years.

Analysis of worldwide projects related to advanced light-water reactors has shown that their designs have tended to become more complicated and their construction increasingly complex. Reducing the probability of severe accidents and the severity of their consequences, such as core melting, historically has led to an increase in the number and complexity of engineered safety systems. However, reducing the probability that severe accidents can occur does not eliminate them completely. If it is assumed that the probability of a large fission product release from the containment is equal to $\sim 10^{-7}$ per year in the case of one severe accident at one advanced VVER reactor. The total worldwide probability of a large fission product release from containment can be easily estimated under the condition that world energy consumption is raised to the level of developed countries (per capita) if all of the required additional energy is to be provided only using VVER nuclear powerplants. Under the previous assumptions, the total number of nuclear reactors has to be equal to $\sim 10^4$ worldwide and the estimated number of severe accidents leading to a fission product release will be equal to $10^4 \times 10^{-7} \sim 10^{-3}$ per year. This estimate is very high and will not be accepted by the population.

In addition, extensive worldwide deployment of the current types of fast breeder reactors and pressurized water reactors will lead to the quick transformation of expensive materials, such as zirconium, nickel, chromium, and titanium, for which resources are very limited, to radioactive wastes without the chance to consider other applications for them. Therefore in the future, the worldwide development of nuclear energy technology will be possible only if the technology will be based on nuclear reactors which prevent severe consequences under any accidental event. Furthermore, they will have to utilize only inexpensive materials which are available worldwide in large amounts. It seems that high-temperature gas cooled nuclear reactors with fuel in the form of small particles coated with pyro-carbide (microparticle fuel) may be able to satisfy these requirements. Other special features of these nuclear reactors are a very flexible fuel cycle and good neutron economy.

At present time, sufficient technology for the fabrication of micro-particle fuel kernels has been developed and demonstrated on a practical scale. Micro-particle fuel can be obtained with kernels made from solutions of any nuclear fuel materials, ie., uranium, plutonium, thorium and a mixture of these materials. Technologies for micro-particle fuel production based on plutonium and a mixture of uranium and plutonium, have been successfully applied at a Siberian chemical plant in Russia.

As already noted, a plutonium stockpile of ~100 tons of plutonium is not sufficient to solve the problem of long-term energy demand through its use as fuel in nuclear reactors. This leads to a simple question: Is it really desirable to plan for further plutonium production in order to achieve a closed fuel cycle for the Russian nuclear power industry several decades in the future?

As can be concluded from the estimates already presented, it seems reasonable to utilize a technology, which is based on fuel prepared from pure plutonium or MOX in the form of micro-spheres. The project for the development of the GT-MHR reactor, which is proposed as a micro-fuel plutonium burner, is very well known. It may be useful to recount the weak points of this project as they were determined by specialists from the (US) DOE.

Utilization of plutonium in the form of micro-particle fuel for one nuclear reactor can be considered as a good demonstration for this new technology, but it is important to have a great amount of experience in order to depend confidently on this strategy for disposing of plutonium and for meeting Russia's energy needs. In the GT-MHR powerplant concept, the system which accomplishes the energy conversion to electricity employs an electrical generator, turbine, and compressor which are situated on a common heavy shaft and this shaft has one common support system based on magnetic bearings. Such a system has never been tested as one unit. Application of such a complicated electrical generator, turbine, and compressor system will be even more difficult in the case of its incorporation with a nuclear reactor. Both of these arguments are heavy enough to support the conclusion that a very long period will probably be required for the development of this reactor, and its energy conversion system components. In addition, it may be necessary to commit financial resources that are much higher than the current estimates.

However, ways to circumvent the above mentioned difficulties already exist. Reduction of the maximum temperature on the working surfaces of the fuel elements and the use of coatings on the microparticles may reduce the probability that a failure of these surfaces will lead to radioactivity leaks. Also, it should be noted that it is possible to substitute a more technologically conservative design for the energy conversion system in order to increase reliability and decrease development time. In any case, nuclear reactors, which contain only fissionable materials and graphite in their cores, will have very good neutron economy if their physical sizes are not too small. This fact supports the concept of GT-MHR reactors using near-breeding thorium based fuel cycles. Such reactors will have an effective neutron multiplication factor close to unity even at high levels of burnup. If pure plutonium is used as a fuel and thorium is loaded separately into separate assemblies, very high levels of plutonium burnup can be reached. To accomplish high amounts of plutonium destruction, the reactor core needs to be operated with infinite neutron multiplication factor in the plutonium fuel that is significantly less than unity during the last part of the plutonium fuel's lifetime in the reactor.

5. Options for Pu Disposition

Operation of any radiochemical processing plant will lead to a non-zero plutonium release to the environment and the possibility for its spreading over the earth. The values for specific plutonium releases are the highest in the world for the current Russian radiochemical technology. This means that in Russia, more than anywhere else, radiochemical fuel reprocessing has to be kept to a minimum after fuel irradiation. Therefore, three options for the management of excess weapons-grade plutonium and reactor-grade plutonium are proposed in the case that plutonium utilization (and dispositioning) through its utilization as a fuel for nuclear reactors is adopted.

The first option consists of accomplishing the following tasks:

- design and construction of facilities which are able to process and fabricate MOX fuel for pressurized water reactors;
- radiochemical reprocessing of MOX fuel after irradiation with remote handling of not only the plutonium, but also the uranium because the uranium will be contaminated with ^{232}U;
- conversion of existing light-water reactors to operate with MOX fuel also has to be accomplished.

In the case of that this option is realized, maximum destruction of the excess plutonium in the near term will be achieved. The advantages of this option are as follows:

- existing light-water reactors can be modified to use MOX fuel without great difficulty;
- there already exists a wealth of worldwide operational experience with MOX fuel in LWRs.

Disadvantages of this option are as follows:

- it makes radiochemical processing of spent nuclear fuel unavoidable which (especially in Russia) may lead to a considerable potential for significant plutonium releases to the environment;
- it requires considerable amounts of capital investment in order to construct and operate the new MOX fuel production facilities and additional radiochemical processing plants.

The second option for plutonium utilization (and dispositioning) through its utilization as a fuel in nuclear reactors is very similar to the first one, but the plutonium is projected to be utilized as a fuel for fast reactors.

The advantage of this option is that

22

- production of MOX fuel for fast reactors from weapons-grade plutonium (and from pure uranium) have already been accomplished on a small scale.

The disadvantage of this option is that:

- it requires a considerable amount of capital investment in order to construct fast reactors, fuel production plants, and radiochemical processing plants based on fully automated technology.

In the case of the first two options, the amount of plutonium, which is committed to supplying nuclear power production, will necessarily need to be increased every year in order to meet demand. In addition, supplies of expensive materials, such as zirconium, nickel, chromium, and titanium, which are very limited will be depleted.

If the first option is implemented, the potential energy resource available from uranium can be doubled. If the second option is implemented, the potential energy resource available from uranium can be increased by a factor of 3 or 4. After that, the fraction of ^{232}U in the uranium will have risen increased to a level which is difficult to predict. The increased fraction of ^{232}U will eventually require new fuel processing facilities based on fully remote handling technology and the consequences of radioactive releases resulting from severe accidents will increase. The first and second options may be implemented together in a wide range of proportions.

The third option consists of developing and deploying nuclear reactors with fuel based on micro-particles coated with pyro-carbide layers.

The advantages of this option are as follows:

- No need for capital investment in radiochemical processing facilities and the capital investment for the fuel production technology is much less than in other options;
- Eliminates the possibility of plutonium releases to the environment from radiochemical processing facilities.

Disadvantages:

- Currently, all of the nuclear reactors based on helium cooling technology are shut-down or have been permanently dismantled;
- GT-MHR system appears to be complicated, uses some large untested systems and has yet to be proven to be inherently safe.

A comparative analysis of the above options for the management of plutonium shows the following advantages and disadvantages. Implementation of the first two options requires increased production and consumption of plutonium and favorable economic conditions and governmental policy for the development and deployment of nuclear energy technology. The probability of plutonium releases from the radiochemical processing plants can be calculated and the effects of releases to the environment can be estimated in future decades. The second option requires the highest amount of capital investment per unit of produced energy because this option calls for construction of both new reactors and new radiochemical processing facilities to achieve its efficient fuel resource utilization with a high level of safety. The third option appears to require the smallest commitment of financial resources, but the technological nature of the reactors is very advanced and, perhaps, will require large investments of capital for development.

The third option may not lead to an increase in the amount of plutonium in existence because plutonium is only used as startup fuel for initiating a thorium based fuel cycle which depends on ^{233}U as the fissile isotope. Within this scenario, separated plutonium can be used up completely. The third option also does not call for depleting precious stocks of scarce materials and it yields a minimal amount of radioactive wastes. Nowadays, the available amount of cheap thorium is incredibly high in the form of refined concentrate and it will not require additional investments for its mining in the foreseable future.

Thus, the main choices related to the further exploitation of nuclear energy technology to meet Russia's future energy requirements are between full conversion to a "Plutonium Age" or rejection of it. But even in the case of complete rejection of a "Plutonium Age", plutonium may, nonetheless, play an important role by providing startup material for a nuclear age fueled with thorium and ^{233}U.

It should also be noted that the advantages of microparticle fuel utilization may not as yet be perceived completely. Particularly, in some proposed nuclear powerplant designs, this fuel may make it possible to employ a very low pressure in the primary coolant circuit while maintaining an already achieved typical temperature range in the microparticle fuel of ~ 1000° to 1200° C. Currently, the possibility of developing and deploying such a system are being studied in Russia. Technologies using core materials that will allow heat conduction to be a more important mechanism for core cooling are being investigated in experiments. In the case that these technologies are successfully developed, the strength requirements on the reactor pressure vessel may be reduced significantly, and hence, may make reactor construction much more simple and safe. As a result, the customary schemes for energy conversion, which were proven through industrial applications some time ago, can be considered instead of the GT-MHR energy conversion scheme.

THE ROLE OF NUCLEAR POWER IN SUSTAINABLE DEVELOPMENT

V. MOUROGOV
V. KAGRAMANIAN
M. RAO
International Atomic Energy Agency
Wagramer Strasse 5,
PO Box 100
A-1400 Vienna
Austria

1. Introduction

At the United Nations Conference on the Environment and Development held in Rio de Janeiro in 1992, governments adopted Agenda 21 - A programme of Action for Sustainable Development Worldwide in 21st Century, the Rio Declaration on Environment and Development. In dealing with energy, the Agenda 21 message was unambiguous: "energy is essential to economic and social development and improved quality of life. Much of the world's energy, however, is currently produced and consumed in ways that could not be sustained, if technology were to remain constant and if overall quantities were to increase substantially". What is the present status and what role might nuclear power play in addressing the energy challenge? What are the IAEA activities in this field? These are the two topics discussed in this paper.

2. The Energy Challenge

2.1 AGENDA 21 AND ENERGY

During the past century, energy has, brought vast benefits to the industrializing world. Energy will continue to play a principal role in promoting economic growth and improving human well-being. The coming decades will see global energy consumption increasing substantially, driven by economic development and population growth in the developing world. Today's developing countries with some three-quarters of the world's inhabitants consume only one-fourth of global energy.

T. A. Parish et al. (eds.),
Safety Issues Associated with Plutonium Involvement in the Nuclear Fuel Cycle, 25–34.
© 1999 *Kluwer Academic Publishers. Printed in the Netherlands.*

26

A 1995 study by the World Energy Council (WEC) and the International Institute for Applied System Analysis (IIASA) projects a range of energy demand increases from 50% for the ecologically driven case to 250% for the high economic growth case, with the latter reaching a 50% increase as early as 2020. The United States Department of Energy in its recently released International Energy Outlook (1997) projects a 54% rise in global energy demand as early as 2015. About half of this increase is due to rising demand in the newly emerging Asian economies, including China and India.

At present, fossil fuels provide 87% of commercial primary energy globally. Nuclear power and hydroelectric each contribute 6%. The non-hydroelectric renewables - solar, wind, geothermal and biomass - that are subsequently referred to as renewables, constitute less than 1% of the energy supply. Fossil fuels also play a dominating role (63%) in electricity generation accounting for about one-third of primary energy consumption. Nuclear and hydroelectric sources provide 19% and 17%, respectively, with renewable systems accounting for less than 1%.

There has been some progress in reducing environmental pollution , particularly noxious gas and toxic substance emissions from fossil plants through costly pollution abatement technologies, such as, desulfurizers, nitrous oxide reducers and precipitators. Globally, however, there remain serious environmental and health impacts due to persistent releases. Pollution in today's developing countries with their heavy reliance on fossil fuels and the absence of abatement technologies, is reaching destructive levels, particularly in urban areas. Globally, the large quantities of waste containing toxic pollutants, particularly from coal combustion, pose a long term problem for water and food chains. A single large 1000 MWe coal plant annually produces large quantities of waste, i.e., some 320,000 tonnes of ash containing 400 tones of heavy metals like arsenic, cadmium, cobalt, lead, mercury, nickel and vanadium.

There has been little progress in reducing the fossil fuel greenhouse gas (GHG) emissions that are projected to lead to atmospheric warming leading to global and regional climate change. Carbon dioxide (CO_2) and methane (CH_4) are the principal greenhouse gases arising from human activities. Global energy services accounts for more than half of man-made greenhouse gas emissions. There is no economically viable technology on the horizon to abate the enormous quantities of atmospheric CO_2 emissions.

For many decades fossil fuels will continue to be the major energy source with methane gas becoming a major component. Countries having or exporting fossil fuels cannot easily turn away from these energy sources and likewise the economically dynamic countries of Asia cannot easily turn away from fossil fuels to uncertain and currently costly renewables.

The Intergovernmental Panel on Climate Change (IPCC), a scientific body established in 1988 by the World Meteorological Organization and the United Nations Environmental Programme to deal with climate change, reported in their 1995 Second Assessment Report, that the continuation of current global levels of GHG emissions might cause climate change by 2100 with significant environmental consequences including effects on ocean levels and regional precipitation resulting in a wide range of social and economic impacts.

Scientists agree that global warming and the resulting climate disruptions could seriously harm human health; increase the incidence and intensity of floods and droughts decreasing food production in some of the world's poorest nations; and threaten the survival of many plant and animal species. Climate change concerns everyone. We all know what climate we have, we do not know what climate we might get.

It is evident that balancing the required energy to fuel social and economic progress in the 21st century with environmental needs will be no small challenge. The significance of this challenge led the world's leading economic countries and Russia at their July 1997 Denver Summit to call for a ministerial meeting on energy issues in Moscow in early 1998 with a report of its result to be available for their mid-year Summit.

Minimizing the impact of possible global climate change has become one of the principal goals of the sustainable development movement. At present there is much talk about the need to reduce GHG emissions. However, neither the extent of global climate change nor its potential impact, especially at the regional level, is yet fully understood. Because of this uncertainty, many advocate that the world should pursue so-called "no regret" strategies, by which they mean energy policies which society would not regret even if the fear of global climate change proves unfounded. In essence, no-regret strategies involve the re-orientation of the world's energy system away from the current extensive use of carbon fuels, and is often referred to as the "decarbonization" of the energy system.

2.2. NUCLEAR POWER FACTS

The Second Assessment Report of the IPCC (1995) lists four major options for the decarbonization of the world's energy system, with quasi-zero, or at least without significant cost penalties:

1. More efficient use of energy;
2. Switching to modern renewable sources of energy (wind, solar and biomass);
3. More use of nuclear energy; and
4. More efficient conversion of fossil fuels through an accelerated use of the next generation of clean fossil technologies including decarbonization of flue gases and fuels, and CO_2 storage.

Meanwhile, it appears that nuclear energy is increasingly discounted as a viable and no-regret GHG mitigation option at many international forums on the world energy future. For example, during the 1997 UN General Assembly's deliberations on Sustainable Development, nuclear power was essentially written off.

One of the main arguments that is often used against nuclear power is public concerns about reactor *safety*, nuclear weapons *proliferation*, the transport and disposal of radioactive *waste,* and *economics.* Although current public concerns are a crucial argument and should be addressed in very serious ways, nevertheless to write off nuclear power would be environmentally and economically imprudent.

While energy efficiency - in generation, transmission and end-use - and new technologies are a necessary element of sustainable energy policies, it should be understood that they will be far from adequate to compensate for the expected increase in energy demand. According to World Energy Council estimates, the share of renewables could increase from the present 1% to 5-8% of the total commercial and non-commercial energy supply projected for 2020. But it is hard to imagine that these sources could be economically viable for large-scale base load electricity generation. The intermittent character of solar and wind energy is also likely to be a severe handicap so long as we do not have better means of storing electric energy.

In fact nuclear power is the only fully developed non-fossil electricity generating option with potential for large-scale expansion. In case nuclear power is limited to the present level or even less due to political reasons, there will certainly be a need for increased use of fossil fuel for base load electricity generation, and in many cases, this would mean increased use of coal.

The challenges for realizing the necessary revival of the nuclear option lie, 1) in improving the technical and economic performance of nuclear power plants while enhancing even further their safety, 2) satisfactorily addressing the practical issues of waste management and disposal and 3) in providing objective, authoritative, reliable and reproducible information to correct public perceptions.

Nuclear power safety. The objections to the use of nuclear power on the grounds of safety may gradually be answered by positive experience. No accidents in the world have had more publicity than those at Three Mile Island and at Chernobyl. This has tended to overshadow the fact that by now the world has the experience of some 8,100 reactor years of operation without any other major accident. Through national regulatory organizations, the World Association of Nuclear Operators and the IAEA the lessons of these many years of experience are made available to all operators.

The Three Mile Island accident in 1979, even though it did not spread any radioactivity into the environment, triggered extensive safety reviews, strengthening nuclear safety in the Western world. The Chernobyl accident, which occurred 11 years ago, similarly led to reviews and new safety measures in Russia and Eastern Europe. Thus these two major nuclear accidents, which provoked so much opposition to nuclear power, also set in motion determined and extensive action in the field of safety - at the national and international level, at the design level and at the operations level.

Nuclear safety became even more of an important international concern and the IAEA became a central instrument through which governments co-operate to establish important elements of what is now termed an "international nuclear safety culture". The impact of this effort can be seen in the improved production figures for nuclear power plants around the world, lower doses to their personnel and fewer unplanned stoppages. New types of advanced reactors, some of them available in the market today, have new safety features and can be expected to have even better records for reliability and safety than the current reactor types.

Nuclear waste disposal. The final disposal of high level radioactive waste is technically feasible but still needs to be demonstrated convincingly to the public. That this has not been done is largely attributable to public skepticism or opposition and lack of the necessary political support. Therefore, presently, high level wastes are being stored above or below ground, awaiting policy decisions on their long-term disposal which will have to materialize at some point. Modern radioactive waste disposal concepts satisfy very high demands for safety and are vastly preferable to the ways we deal with the wastes originating from fossil fuels and other chemical and manufacturing sources.

Non proliferation issues. The concerns that an expansion of nuclear power might lead to a further spread of nuclear weapons and to illicit trafficking in nuclear materials, are also not to be ignored. However, it is worth recalling that nuclear weapons development consistently preceded - and did not follow from - the introduction of nuclear power reactors. To eliminate the production or diversion of weapons-grade materials, the permanent Treaty on the Non-Proliferation of Nuclear Weapons (NPT) of 1970 commits 180 countries to refrain from acquiring nuclear weapons and to accept comprehensive IAEA safeguards on all their nuclear activities. Also, the IAEA's safeguards inspection system has recently been strengthened, regional Nuclear-Weapon Free Zones exist in most areas of the world, nuclear testing has ended, and the trend towards nuclear disarmament in the post-Cold War era is manifest. The physical protection of nuclear materials to help in combating possible illegal trafficking attempts is also being reinforced.

Economic compatibility. The cost of energy production remains important to countries, utilities and consumers. After several decades of development by governments and investment by electric utilities in many countries, nuclear power is a commercially proven energy generation option. Nuclear power is, at present, roughly on par with coal in most regions, and in many cases it is also competitive with natural gas in countries without a natural gas infrastructure in place. However, high up-front capital costs and long amortization periods could be a barrier to the large-scale deployment of nuclear in capital-starved developing countries. This may become even more of an obstacle due to electricity market deregulation and privatization. Therefore, the nuclear industry is challenged to offer new types of reactors with reduced construction times and lower capital costs. Indeed, several such concepts are under development.

Of course, even for the advanced nuclear plants, their capital cost will be higher than the capital cost of fossil plants. But the important factor is the very low cost of nuclear fuel. Uranium resources are plentiful and have no other use. Thus, once built, nuclear power plants produce electricity at a cost that is relatively insensitive to inflation or the fluctuations of prices on the world energy market. Reactors now tend increasingly to stay in operation past the end of their depreciated economic life of about 30 years, further increasing their profitability.

Even if the prices of fossil fuels do not increase over time, imposition of a carbon tax or external costs on energy generation and use, would change the overall cost economics of electricity generation very much in favor of nuclear power. Indirect costs, such as those for waste management and decommissioning, are already components of nuclear power generation costs. New nuclear plants will incorporate greater economic advantages in addition to higher availability and improved safety features. Nuclear electricity generation can thus be expected to keep its competitive edge in various countries over the long run.

2.3 NUCLEAR POWER ADVANTAGES

It has to be noted that there is no energy source that does not adversely affect the environment, is absolutely risk-free, safe, secure and reliable, and at the same time maximizes economic efficiency. Nuclear power, as all other electricity generating options, has its advantages and disadvantages. Only on the basis of a detailed and comprehensive comparative assessment of the alternative energy options one can develop a national energy strategy which is environmentally and economically sound. To assist energy planners in the member states, the IAEA has, over the years, carried out comparative assessments of energy sources, known as DECADES project that cover a broad range of technical, economic, environmental and health aspects. Extensive databases and analytical tools allow full energy chain analysis.

Limited environmental impacts. Studies of fossil fuels, nuclear power and renewable energy sources show that nuclear power under normal operation is benign to the atmosphere and to the earth and its inhabitants locally, regionally and globally. Owing principally to the small fuel requirements, there are limited environmental impacts for the full energy chain, from mining to waste disposal and decommissioning.

Security of energy supply. Where indigenous fossil fuel resources are lacking, nuclear power can contribute substantially to security of supply. For countries without oil and gas, such as France, Japan and the Republic of Korea, nuclear power offers a measure of self-reliance and immunity against crises. Uneven distribution of fossil energy resources have often been the underlying factor for conflicts between nations and regions. Since nuclear fuel is a highly concentrated energy source, adequate fuel supplies can be procured, transported and stored without any serious difficulties, thus producing economic and political stability in different regions of the world.

The versatility of nuclear power. At present the use of nuclear power is limited to supply of electricity and to such special applications like submarines and ice-breakers propulsion. However, if found preferable to fossil fuels, nuclear power could also be of used to produce heat and steam for industry and domiciles.

Considering the severe shortages of fresh water that are expected in many areas in the world -including coastal areas - nuclear power might also become an important means for desalination of sea water. Among many IAEA members there is growing interest in this topic.

Spin-off benefits. Development of nuclear power technology has had a beneficial effect in shifting a part of the labor force from high risk occupations such as coal mining and transportation to more skilled occupations in industrial, scientific and *R&D* organizations. An impressive nuclear medicine sector has also been built up around the remarkable tracer properties of radioactive materials. Another specific application that is now drawing greater attention is the use of isotope techniques for evaluating human nutritional status and measuring the effects of nutrition programs. Finally, food irradiation has been demonstrated for several years to be completely safe and effective in destroying harmful bacteria that reside in much of our global food supply.

3. Direction Of IAEA Programs Related To Nuclear Power Development

Taking into account the current status of nuclear energy in the world, a stronger initiative on an international level is required to realize the technology's potential benefits. The Agency continues to play a catalytic role in coordinating actions, covering the whole range of energy issues, undertaken by Member States and different international or specialized organizations. The IAEA's programs and activities will be described under the following headings: nuclear power, nuclear fuel cycle, and nuclear energy for sustainable development.

Underlining the work ahead is a reinforced global commitment to safe nuclear operations through legal agreements, common basic safety standards, and associated expert services. The declaration of the April 1996 Moscow Summit reiterated that safety is the first priority in nuclear activities and that international conventions will be major instruments in achieving safety. Furthermore, it is to be expected that safety targets will continue to rise and will require a continuous effort and vigilance by the IAEA and its Member States to ensure that measures to achieve safety levels are implemented , taking into account both the technological and regulatory aspects of the safety problem.

3.1 NUCLEAR POWER PROGRAMME

The IAEA's efforts in nuclear power will focus on the contribution of nuclear energy to sustainable development, with emphasis on:

1. Promoting design and operation measures necessary to achieve safe development of nuclear power;
2. Assisting developing Member States in planning and implementing nuclear power programs and in improving the management of nuclear power projects and operating plants;
3. Improving the operational performance and the reliability of nuclear power plants through sharing of operational experience and information worldwide in all areas, including training and qualification of personnel; and
4. Dissemination of information on advanced as well as innovative nuclear reactor and fuel cycle systems.

One mechanism used by the IAEA to keep abreast of the technological developments in a given area is the constitution of an international working group (IWG) for that area. An IWG consists of top experts from different Member States. The IWG meets periodically to review the current status and future directions of activities in the area concerned and advises the Agency on the programme of activities necessary to meet the needs of Member States.

Through the IWG on advanced reactors (Light Water Reactors, Fast Sodium Cooled Reactors, High Temperature Gas Cooled Reactors, Heavy Water Cooled Reactors) the Agency will encourage an international exchange of information on non-commercial technology and on co-operative research. Another important function will be to assist countries in the preservation of key technological data related to advanced nuclear power systems.

The Agency will also continue to provide a forum for the review of information on the development of innovative nuclear energy systems such as:

1. Small and medium sized nuclear reactors with passive safety features;
2. Thorium fueled reactors;
3. Fast reactors cooled by heavy metal; and
4. Accelerator driven and hybrid fusion/fission concepts.

A new area of Agency activity relates to the current need to examine the possibility civilian use of nuclear technologies developed for naval and space applications. Another area concerns desalination. An important event of 1997 is the International Symposium on Desalination of Sea Water with Nuclear Energy. Results of this symposium will be utilized to define more precisely the IAEA's work in this area.

3.2 NUCLEAR FUEL CYCLE PROGRAMME

Among the key topics addressed in the recent IAEA nuclear fuel cycle symposium were the comparative assessment of different options for development of the fuel cycle, management of spent fuel and plutonium and disposal of radioactive waste.

The volume of spent fuel in interim storage at both power and research reactors is growing, and the long term storage of spent fuel in aging facilities will become an increasingly crucial issue, regardless of the management option chosen. Identification and mitigation of environmental, health and safety vulnerabilities of aging spent fuel will be emphasized and activities relating to exchange of information, experience and advice on technical solutions in this area will be expanded.

The focus of activities relating to radioactive waste management will be on the following:

1. Collection, assessment and exchange of information on waste management strategies and technologies;
2. Provision of general technical guidance, assistance in technology transfer, and promotion of international cooperation; and
3. Examination of the long-term prospects for regional waste management facilities to provide new opportunities to developing countries, particularly small ones without modern fuel cycle infrastructure, in resolving their waste management problems in a cost effective manner.

The Agency's programme concerning plutonium management will reflect the realities facing the international community today, including the security and commercial impacts of ex-weapons material. The activities must also be geared to promoting further the reliability, safety and economic viability of nuclear power. The Agency's role can be generally classified into the following five categories:

1. Verification;
2. Forum for information exchange;
3. Establishment of norms;
4. Roles specified in the statute of the IAEA; and
5. Co-ordination of international efforts on disposition options.

Of course, the traditional verification function will remain an essential part of the IAEA's role. The verification of physical inventory is being carried out by the IAEA in facilities which include storage, research reactors, power reactors, reprocessing plants and fuel fabrication plants. This is a well known subject which is the responsibility of the Safeguards Department of the IAEA.

The classical function of the Agency as a forum for exchange of experience, ideas and approaches through various meetings is very important. A fuel cycle strategy is essential for nuclear energy to play a significant role in providing a sustainable energy source for long term global development. The IAEA, as an autonomous organization under the UN system, can play an important role in assisting Member States in the development of safe, secure and economic management of plutonium, while at the same time ensuring through its safeguards functions that there is no diversion for non-peaceful purposes. The IAEA can provide to interested Member States regularly updated information about projected stockpiles of separated civil plutonium, and can supplement this information with data on the inventories of nuclear materials no longer required for defense purposes.

Using the Agency as a means for elaborating international norms should also be examined. The Agency has already published three safety guides on interim storage of spent fuel from power reactors. They cover the design of spent fuel storage facilities, the preparation of safety analysis reports and the safe operation of spent fuel storage facilities. Preparation of a safety document on the safe handling of plutonium is nearing completion.

The Statute of the IAEA grants broad authority to the IAEA to receive, store, manage and control nuclear material, and in particular, to require the deposit with the Agency of excess special fissionable material recovered or produced as a by-product of non-proscribed uses, in order to prevent stockpiling of such materials. Public concern about the safety, security and proliferation of the stockpile of civil plutonium, both separated and contained in spent fuel, and stockpiles of separated plutonium resulting from dismantled warheads seems to suggest the need to examine the desirability and feasibility of measures designed to address these concerns. Additionally, the involvement of the IAEA in possible international co-operative programs for the disposition of plutonium might help to make such programs more effective and to promote transparency.

In the medium term perspective of the IAEA for 1998-2003, it is proposed to establish (in 1998) an international working group on nuclear fuel cycle as a mechanism for dialogue among Member States on plutonium and related fuel cycle issues. The mission of the Working Group is to provide support for the development and improvement of nuclear fuel cycle operations with a view to enhancing safety and reducing the risk of proliferation of nuclear weapons and to search for and elaborate nuclear fuel cycle strategies for the use and disposition of spent fuel and plutonium.

3.3 NUCLEAR ENERGY FOR SUSTAINABLE DEVELOPMENT

Economic and environmentally sound energy systems will be needed to meet the increasing energy demands for sustainable development. Nuclear energy has the potential to play an important role in the future energy mix in different regions of the world. The objective of the IAEA programme on comparative assessment of sources of energy is to define optimal strategies for the development of the energy sector, consistent with the aims of sustainable development. The programme will focus on:

1. Comparative assessment of economic, health, and environmental aspects of energy systems and introduction of the results into the process of energy policy formulation and electricity system expansion planning;
2. Enhancement of the capability of Member States to incorporate health and environmental considerations in the decision making process in the energy sector; and
3. Provision of a basis to define optimal strategies for the development of the energy sector, consistent with the aims of sustainable development.

A key element is the development and dissemination of databases and methodologies for comparative assessment of energy sources in terms of their economic, health and environmental impacts. Consideration will also be given to dealing with energy demand and supply issues outside the electricity sector.

4. Conclusions

Continuing the current dependence on fossil fuels is not sustainable. Nuclear power can play a role in mitigating the detrimental environmental impacts of energy use. To make nuclear energy more attractive, measures to gain public acceptance will be necessary. The adequacy of waste management policies and the disposal of high-level waste will be demonstrated through selection and use of geologically acceptable depositories. To maintain and enhance nuclear power's performance and safety record, there will need to be continued vigilance to improve design, to implement an effective operational safety culture, and international safety agreements.

The IAEA will have to play an increasingly important role in coordinating the efforts of Member States and other international organizations in order to realize the potential benefits of nuclear energy for global sustainable development. An important element of these programs will be improving regional and international co-operation and sharing of infrastructure facilities, development costs, and operational experience to sustain the development of nuclear technology in a safe, reliable, and economic manner.

PLUTONIUM DISPOSITION RESEARCH AND RELATED ACTIVITIES AT THE AMARILLO NATIONAL RESOURCE CENTER FOR PLUTONIUM

D. R. BOYLE
Department of Nuclear Engineering
Texas A&M University
College Station, TX 77843-3133
United States

R. S. HARTLEY
Technical Director
Amarillo National Resource
Center for Plutonium
Amarillo, TX 87979-7797
United States

1. Introduction

With the end of the Cold War, the United States and Russia are reducing their nuclear weapons stockpiles. Deciding what to do with the materials from the thousands of retired nuclear weapons is an important international challenge. To help address this question and related issues, the US Department of Energy (DOE) and the State of Texas established the Amarillo National Resource Center for Plutonium in 1994.

The Center functions as a scientific and technical resource on issues related to nuclear weapons materials, including plutonium disposition and storage, environment, health and safety, and nonproliferation. The three major universities in the state of Texas, the Texas A&M University System, Texas Tech University, and the University of Texas System, formed a unique research consortium to operate the Center. The Center is headquartered in Amarillo, Texas, home of the US Department of Energy's Pantex Facility, which assembles and disassembles all US nuclear warheads. The three universities have separately conducted research related to Pantex operations for years, but the Center now integrates and focuses those efforts in the Amarillo area, giving area citizens a larger voice in their future. The Center is funded by DOE through a cooperative agreement with the State of Texas.

35

T. A. Parish et al. (eds.),
Safety Issues Associated with Plutonium Involvement in the Nuclear Fuel Cycle, 35–38.
© 1999 *Kluwer Academic Publishers. Printed in the Netherlands.*

The Center directs three major programs that address the key aspects of the Pu disposition issue:

- the Communications, Education, and Training Program, which focuses on informing the public;
- the Environmental, Safety, and Health (ES&H) Program, which investigates the key ES&H impacts of activities related to the DOE weapons complex in Texas; and
- the Nuclear Program, which is aimed at maximizing safety and proliferation resistance by helping to develop and advocate safe stewardship, storage, and disposition of nuclear weapons materials.

This paper provides a top-level overview of the Center's current and planned Nuclear Program research activities. In addition, the authors identify an important contribution that this unique type of organization makes to open and informed public debate on nuclear weapons-related issues. The Center's Nuclear Program activities are described below under three broad categories: international activities, storage, and disposition.

2. International Activities

One of the key goals of the US effort in nuclear materials disposition is to develop a parallel capability and disposition program in Russia. Through interactions with counterpart organizations, the Center seeks to expedite the safe storage and disposition of nuclear materials in Russia simultaneously with the process in the US.

The Center's original efforts to initiate interest in joint US-Russian research involved several Phase I projects, which included investigating the use of fast reactors (in Russia) to burn excess weapons-grade plutonium, and developing cooperative efforts, focused on nuclear materials management, with institutions of higher education in Russia and elsewhere in the Former Soviet Union. These and other Phase I activities culminated in the publication of the "Joint United States/Russian Plutonium Disposition Study" in September, 1996.

As an outgrowth of these Phase I activities, the US - Russian Steering Group met in Washington DC in May, 1996 and signed a protocol initiating ten joint US-Russian research efforts. The Center has responsibility for six of these efforts:

- Validation of Russian LWR Reactor Design Analysis - Validates reactor computer codes for weapons-grade MOX use in Russian VVER-1000 reactors to allow optimization of fuel loads.
- Russian LWR MOX Fuel Tests - Fabricates VVER-1000 MOX fuel from weapons-grade (W-G) plutonium for future reactor burnup tests to determine the feasibility of substituting W-G MOX for slightly enriched uranium fuels.
- CANDU MOX Parallax Fuel Tests - Fabricates CANDU MOX fuel for shipment to Canada's Chalk River reactor for future reactor burnup tests to

determine the feasibility of substituting W-G MOX for natural uranium fuels to allow burnup of excess weapons-grade plutonium in a non-participating country. This activity is currently in a "hold" status.

- Large Scale Tests of Plutonium Immobilization - Performs tests to demonstrate the immobilization of plutonium in glass for long-term disposition in geologic depositories.
- Geologic Disposition - Performs experimental tests and analysis of plutonium sorption in rock in geologic depositories to aid in the design of future geologic depositories.
- Analysis of Russian BN-600 Reactor for Plutonium Burning - Uses US safety analysis computer codes to aid in the conversion of the BN-600 Russian Reactor to burn existing excess W-G plutonium as opposed to breeding more plutonium.

As international relations progress, the Center plans to continue working with the Russians in these and other disposition efforts.

3. Nuclear Weapons Materials Storage

The Center focuses in the storage area on scientific and technical issues associated with intermediate storage of nuclear weapons materials and components. The goals of this effort are to enhance worker safety, reduce costs, and ensure the reliability and effectiveness of reusable components. The specific areas currently under investigation include:

- Facility Design - The Center aids in the design of storage facilities by addressing issues related to criticality safety, shielding, radiation safety, thermal analysis, robotics, material control and accountability, and container surveillance.
- Facility Analysis - The Center is currently conducting two projects aimed at ensuring existing storage facilities are capable of addressing current and future storage missions. The first project addresses aircraft accident forecasting to assess the risk of long-term plutonium storage scenarios. The second effort models the transport of potentially toxic aerosols, following a hypothetical chemical explosion, to determine the effectiveness of engineered barriers in the work area.
- Container Evaluation - The Center is participating with the national laboratories in the design of new storage and shipping containers and evaluating the life-limiting aspects of existing storage methods. This work includes the assessment of radiation damage and microstructural changes to stainless steel in contact with plutonium, the development of novel storage container monitoring techniques, and improvements to the sampling systems for the detection of plutonium. In addition, the Center plans to

38

develop a suite of analytic tools to enable it to rapidly evaluate all key performance characteristics of proposed new storage container concepts.

4. Nuclear Weapons Materials Disposition

In the nuclear weapon materials disposition area, the Center investigates the scientific and technical issues associated with the different options for Pu disposition following weapon disassembly. The objective is to maximize the use of existing nuclear resources while guarding against the proliferation and reuse of these materials in military applications. Specific efforts being pursued include:

- Plutonium Conversion - To aid in overcoming the obstacles involved with using weapons-grade plutonium as reactor fuel, the Center is investigating processes to convert plutonium metal to plutonium oxide. Currently, Center researchers are developing new methods for gallium removal and experimentally measuring the impact of gallium-cladding interactions at prototypic reactor temperatures.
- Water Reactor Options - The Center is assisting the DOE by leading an extensive computational effort in support of the national program to burn MOX fuel, made from weapons-grade plutonium, in US light water reactors. This project includes benchmarking reactor design codes, generating a MOX data repository, and modeling the VVER-1000 power plant.
- Non-Proliferation and Transportation - The Center is working to minimize the proliferation threat by conducting risk analyses of the alternative means under consideration for transporting plutonium oxide and MOX fuel within the US.

5. Public Opinion

One aspect of the Center deserves special mention because of its potential value to like-minded organizations throughout the world. The Center's unique structure, its mode of operation, and its carefully crafted management priorities allow it to function as a highly effective and trusted source of information for the public. University researchers (professors and students) conduct the scientific investigations listed above, frequently collaborating with their colleagues in Russian universities and in US national laboratories. All these investigations are accomplished in a fully open, publicly accessible fashion. The Center's Communication/Education group works hard to ensure the results of these scientific projects are widely shared with the local community through a variety of mechanisms. As a result, the local Amarillo citizens have developed an unusually thorough and factual knowledge base from which to make informed decisions about the necessary, post-Cold War changes proposed for Pantex operations. At a time when government-operated nuclear facilities frequently generate widespread skepticism, both the US DOE and the citizens of Texas benefit significantly from the existence of a knowledgeable and generally supportive public attitude.

THE ISTC PROJECTS RELATED TO PLUTONIUM UTILIZATION AND DISPOSITION (OVERVIEW)

A. GERARD
L.V. TOCHENIY
International Science and Technology Center
Luganskaya ulitsa, 9
PO Box 25
115516 Moscow,
Russian Federation

1. ISTC - History, Current Status, and Prospects

The ISTC is operating under the auspices of an intergovernmental agreement between the Russian Federation, European Union, Japan and the United States.

The Center started operation on March 2nd, 1994 and since then Finland and Sweden have acceded as funding parties. Commonwealth of Independent States (CIS) representation has expanded to include Georgia, Belarus, Armenia, Kazakhstan and Kyrgyzia. All of the work of the ISTC is aimed at goals defined in the ISTC Agreement:

- To give CIS weapons scientists, particularly those who possess knowledge and skills related to weapons of mass destruction and their delivery systems, opportunities to redirect their talents to peaceful activities;
- To contribute to solving national and international technical problems;
- To support the transition to market-based economies;
- To support basic and applied research; and
- To help integrate CIS weapons scientists into the international scientific community.

The ISTC engages in a variety of activities aimed at meeting these goals:

1. Providing support for scientific research and development;
2. Stimulating collaboration of CIS institutes with foreign institutes, industries and universities. The Parties make great efforts to seek

T. A. Parish et al. (eds.),
Safety Issues Associated with Plutonium Involvement in the Nuclear Fuel Cycle, 39–44.
© 1999 *Kluwer Academic Publishers. Printed in the Netherlands.*

foreign partners that are interested in the proposed projects, to support interaction of the project participants with foreign collaborators and to facilitate participation in scientific conferences.

3. Organizing of the ISTC Seminar Program on subjects of national and international interest. Projects may be funded through both governmental funds of the funding parties specified for the ISTC, and by organizations, nominated as funding partners of ISTC. Funds used within ISTC projects are exempt from CIS taxes.

Projects range from solving environmental problems related to nuclear industries and nuclear safety to the development of new vaccines for contagious diseases. As of July 1997 more than twelve hundred proposals had been submitted to the ISTC, of which 450 were approved for funding, for a total of approximately US$ 145 million. The number of scientists and engineers participating in the projects numbers about 20000. As of June 1997, the ISTC had received more than 40 proposals at various stages of research and development related to the important area of plutonium disposition. More than 25 Russian and CIS institutions are involved in these proposals. The most active institutions are shown in Table 1.

TABLE 1. The most active institutes in submitting and participating in the plutonium disposition projects

Technical area	Institutes	Number of projects
Reactor Concepts	IPPE, OKBM, NIKIET (ENTEK), ITEP	7
Modeling and Experiments	IPPE, MIFI, NIKIET (ENTEK), VNIIEF, KIAM	4
Nuclear Data	IPPE, MIFI, RI (Khlopin)	4
Fuel Cycle	IPPE, VNIINM (Bochvar), NIIAR (Dimitrovgrad)	5
Materials, Reactor Fuels	IPPE, NIIAR (Dimitrovgrad), NIKIET (ENTEK), VNIINM (Bochvar)	2
Radioactive waste	VNIIM(Bochvar), PROMTECHNOLOGY, RI (Khlopin)	9

The total number of funded (and submitted for funding) projects related to plutonium disposition is 31. The approved amount of funds for these programs is above

$14 M or approximately 10 % of the total amount of the ISTCs funds. The involvement of the ISTC Parties in the funding of the projects is described in more detail in Table 2.

TABLE 2. The involvement of ISTC Parties in the funding of projects

Country	Number of	funded projects	Total	%
	Funding Party	Co-Funding	Funding, $ M	
EU	5	17	3.8	27
Finland		2	0.3	2
Japan	7	9	5.0	34
Sweden		6	0.5	4
USA	1	8	4.8	33
Total	13	42	14.4	100

2. Plutonium Disposition

As large amounts of plutonium are recovered from retired nuclear weapons, there are continuing discussions and concerns over the disposition of this material in a way that is environmentally responsible while preventing easy reuse in weapons applications. The Russian Government views the retired plutonium as a valuable asset and, as such, is not favorably disposed to burial or underground storage as a permanent solution. The United States Government has recently reversed its position banning the use of plutonium in commercial power reactors and Japan is committed to closing the nuclear fuel cycle and to the construction of fast breeder reactors. Within the European Community, opinion is mixed, with France embracing fast breeder technology, Germany considering the use of MOX in light water reactors, while the remaining nations are not presently engaging in plutonium fuel cycles. All nations are supportive of programs which will enhance the safety and non-proliferation aspects of future plutonium and actinide disposition while promoting environmental responsibility.

In view of the importance of these issues to the world community, ISTC has funded a number of projects which deal with the important issue of weapons grade and civilian plutonium disposition. The projects fall into three general categories, namely:

- Reactors and Nuclear Fuel Cycles,
- Plutonium and Actinide Incineration,
- Long Term Storage of Actinides.

Some representative projects are discussed below.

3. Reactors and Nuclear Fuel Cycles

To make project development more effective, and to encourage cooperation, the ISTC, EU and Russian Institutes have established a Contact Expert Group, which works to coordinate efforts of different projects related to the use of plutonium as reactor fuel so as to avoid duplication. This group also increases the effectiveness of related projects and improves the orientation of new proposals. Participation in this group by other institutes is open upon the approval of the funding parties. A similar coordinating body has been established for a group of projects related to HTGRs.

Several institutes under the leadership of the IPPE (Obninsk) are studying fuel cycle development in Russia, considering the amount of plutonium to be dispositionned, the infrastructure for the nuclear cycle, and the time table for facility start up (project # 369).

The effects of plutonium on the safe operation of PWRs (VVERs), i.e., plutonium fuel's effects on reactivity, burn-up, spectrum, and control systems will be studied at the critical facilities of nuclear weapon and reactor institutes (VNIITF, IPPE, Kurchatov Institute, ENTEK-NIKIET). Analysis will be performed by modeling experimental and calculational benchmarks (projects # 116, 371). Neutron cross-sections and data libraries and calculational models will be verified and evaluated to make computer calculations more exact and reliable.

Besides the problems of manufacturing and using MOX-type plutonium fuel in conventional power reactors (VVERs, BN), the ISTC has projects to develop new reactor concepts both for energy production and for plutonium and minor actinide transmutation (a new fast reactor BN-1300, HTGR High Temperature Gas-cooled Reactor) through collaborations between Russian, western and Japanese institutes (PNC, JAERI, FRAMATOME, General Atomics, Fujie). New types of fuel and their properties are to be studied including metallic types, nitrides, carbides, vibro-packed, dispersion-type, and so on.

The VNIINM (Moscow) and NIIAR (Dimitrovgrad) are working on projects to develop new technologies for various stages of the fuel cycle, including options when weapons plutonium is used (# 173, 272, 273, 279, 534). Experimental techniques for fuel element fabrication are also under development, such as induction slag melting, mill processing, etc. As a result, such fuel should be more attractive technically and the processing technology should be more effective and less threatening environmentally, from a radiation point of view.

An analysis of the radioactive contamination of graphite from the Tomsk-7 plutonium production reactors is planned. This is one of a number of projects which deal with issues related to graphite use in Russian reactors.

Several projects conducted by VNIINM (# 281, 330) are developing new technologies to minimize the mass and hazard of high-level and medium-level radio-active wastes from fuel manufacturing and reprocessing.

4. Plutonium and Actinide Incineration

General aspects of blanket, target and fuel cycle activities for incineration of plutonium, actinides and some fission products using accelerator based neutron sourceshave been developed mainly in Project # 017. This project has been completed and the results have been presented at an International Conference (Sweden, June 1996). Future experiments will study spallation reactions due to intensive proton beam irradiation of solid targets (ITEP, # 157, 477). New proposals in this category which have been submitted to the ISTC, include:

- # 559 - design of a lead-bismuth target for testing at LANL, USA,
- # 442 - conversion of an experimental heavy water reactor into a subcritical unit and study of its control system,
- # 698, # 747, # A-088 - development of molten salt technology,
- # B-070 - modeling of a target-blanket system using a neutron generator as an experimental neutron source.

5. Long-Term Storage of Actinides

Geological and artificial materials are being studied by various institutes for use as barriers for long-term storage of actinides, as matrix materials for radioactive wastes and for actinide immobilization, as components for vitrification processes. Other projects involve the cleaning of actinide residuals from storage pools, decreasing the level of radioactivity of wastes, and methods for decreasing the diffusion of plutonium and radioactive wastes in ground water near burial sites or in sea water, following an accidental release. The goal of these projects is to find environmentally acceptable and reliable methods, materials and technologies for long term underground storage of actinides. Projects # 059, # 307-1, and # 307-2 are in this set.

Project # 332 (VNIIEF, VNIIN) plans to design storage for uranium and plutonium extracted from dismantled warheads. The fissile material will be converted into forms which prevent criticality and which prevent easy conversion to weapons-grade material while being in a form that is suitable for long term storage.

6. Suggestions for Future Activities

1. Develop and validate more precise analytical methods for defining the criticality parameters of different plutonium compositions in storage containers, burial configurations, transport containers, subcritical blankets and experimental installations to assure safe operation. This work will a include review of available and acceptable levels for uncertainties and error correlations.
2. Compare national standards and regulations as they apply to plutonium management and use. This work seeks to formulate an international consensus approach taking all of the national rules and regulations into consideration.

3. Develop and design a transport cask for fuel and other materials containing plutonium.

4. Conduct a feasibility study to identify alternative methods for RBMK spent fuel management. This study would include consideration of long term storage, and reprocessing including plutonium extraction, and take into account future trends for fuel flow through the RT-1 plant.

WASTE PARTITIONING AND TRANSMUTATION AS A MEANS TOWARDS LONG-TERM RISK REDUCTION

E. R. MERZ
Institute of Chemical Technology, Research Center
KFA-Julich
D-52425 Julich
Germany

1. Introduction

The question of safe radioactive waste disposal continues to dominate the nuclear debate in many countries. In addition to the aspect of ensuring nuclear power plant safety, the management of high-level radioactive waste generated in the course of spent fuel reprocessing ranks among the most important factors influencing the further development of nuclear electricity generation.

The management of radioactive waste, especially the long-term disposal of high-level radioactive wastes, is one of the key issues in today's political and public discussions on nuclear energy. Rather than waiting for their radioactive decay, it is in principle, possible to reduce the period of toxicity of the actinides and some long-lived fission products through transmutation of these isotopes in fission reactors or accelerators. It is claimed that the transmutation or "burning" of the actinides and long-lived fission products could reduce their half-lives by two to three orders of magnitude and hence reduce the length of time for which confinement in a repository has to be ensured.

During the late seventies and again for the past four years, some countries (Japan, France, United Kingdom, United States, and to a lesser extent Russia and Germany) have been carrying out various R&D activities dealing with waste partitioning and transmutation (P-T technology) with the aim of reducing the long-term burdens of nuclear waste disposal [1].

Taking present knowledge into account, it must be stated that actinide and long-lived fission product partitioning from high-level radioactive waste and subsequent transmutation utilizing fission reactors or accelerators, cannot be considered as a real alternative to geological disposal, since a safe repository is needed anyway for some of

45

T. A. Parish et al. (eds.),
Safety Issues Associated with Plutonium Involvement in the Nuclear Fuel Cycle, 45–64.
© 1999 *Kluwer Academic Publishers. Printed in the Netherlands.*

the high-level waste. Rather, such activities may be conceived as a research effort to pursue benefits for future generations through long-term basic R&D.

It seems noteworthy that the advocacy of transmutation rules out the currently favored strategy of direct ultimate disposal of spent fuel elements. It must be borne in mind that there is a worldwide consensus to the effect that the present reference nuclear fuel cycle scenario, including deep underground waste disposal in suitable geological formations, gives adequate protection to mankind. According to the International PAGIS project the peak individual risk due to a detrimental health effect, including the normal and all the altered evolution scenarios, is very low and it does not reach 10E-9 per year [2].

However, there is strong interest in seeing whether any further reduction of the long-term potential hazard can be achieved and at what cost. The most serious criticism that can be made of a geological repository program is its limited ability to predict future effects, which can be exemplified by possible changes of chemical properties in the repository, caused by human activities on the ground surface [3].

2. Irradiation Facilities for Transmutation

The nuclear reactions considered for transmutation are induced by neutrons, causing either fission or neutron capture. The reaction products should be either stable or short-lived with decay to stable products to reduce the long-term radiological hazard. In order to obtain an acceptable transmutation rate, high neutron fluxes are required and the transmutation devices with the greatest potential at the present time are high-flux reactors or accelerator-driven neutron sources.

Several transmutation facilities are conceivable [4,5]. In principle, transmutation is achievable in existing nuclear power plants. Thermal reactors are less favorable since the neutron fluxes are not sufficiently high and the build-up of higher actinides is large. A better design is the fast breeder reactor with its hard neutron spectrum suitable for a high yield in transforming the actinides into fission products.

Specially designed reactors are being studied for burning plutonium and the other transuranium elements. The Integral Fast Reactor (IFR) concept developed in the US and the Advance Liquid Metal Reactor (ALMR) studied in Japan are two concepts which aim at the nuclear incineration of transuranics. They are capable of incinerating the actinides at a rate of about 12 % per year. The design of the Advanced Liquid Metal Reactor (ALMR) under development in the USA has progressed sufficiently to provide confidence that this reactor can be available by the year 2005 to serve as a Minor Actinide-Conversion Reactor (MACR).

Better performance is expected from accelerator utilization. The development of high current, high energy proton accelerators has opened up the possibility of using them as high flux neutron sources. Currents on the order of hundreds of mA of protons of about 1.5 GeV are required. With such devices, it would be possible to generate immense amounts of neutrons in a spallation target [6].

Materials undergoing transmutation in the high neutron flux should optimally be continuously circulated through the transmuter blanket using a molten salt medium which allows continuous chemical processing yielding satisfactory decontamination factors. A combination of a subcritical transmutation system and an accelerator-driven neutron source may become self-supporting in electrical energy, or even produce excess electrical energy.

Much R&D work is still required in the field of efficient transmutation machines. The technology required is today merely at the edge of feasibility.

3. The Problem Radionuclides

The long-lived species of concern are among the category of elements which are higher in atomic number than uranium they are called transuranic elements. Of main importance are Np, Pu, Am, Cm and possibly Cf. Fission products of most concern are Tc-99 and I-129. The other fission products sometimes mentioned due to their half-lives of approximately 30 years, Sr-90 and Cs-137, do not really represent a serious long-term problem, also the nuclides C-14, Cl-36, Zr-93 and Cs-135 may be disregarded from the standpoint of inferior specific radioactivity and/or low radiotoxicity [7,8].

The nuclides which are of major concern in the context of long-term safety may depend on the nature and location of the waste repository. There is no doubt that the principal radioelement Pu-239 can be recycled in various reactor types, e.g. Light Water Reactors (LWRs), as Mixed Oxide (MOX) fuel or as fuel in Fast Breeder Reactors (FBRs). There is also consent based upon dose effect calculations that the most dominant long-term hazardous waste nuclides are Np-237 and its precursors Cm-245, Am-241 and Pu-241 followed by Tc-99 and I-129.

One has to bear in mind:

- All of the transuranic elements are fissionable with high energy neutrons (fissible) and some of them are also fissionable with thermal neutrons (fissile)
- Long-lived fission products are not fissionable with neutrons; they can only be transmuted to stable or shorter-lived nuclides by changing the mass number and/or the nuclear charge.

When actinide elements are irradiated by neutrons they either fission or transmute into a higher isotope. Actinide conversion in LMRs is more efficient than in LWRs because the ratio of fission rate to capture rate is higher in LMRs. Figure 1 shows the decay and neutron capture chains for actinides.

48

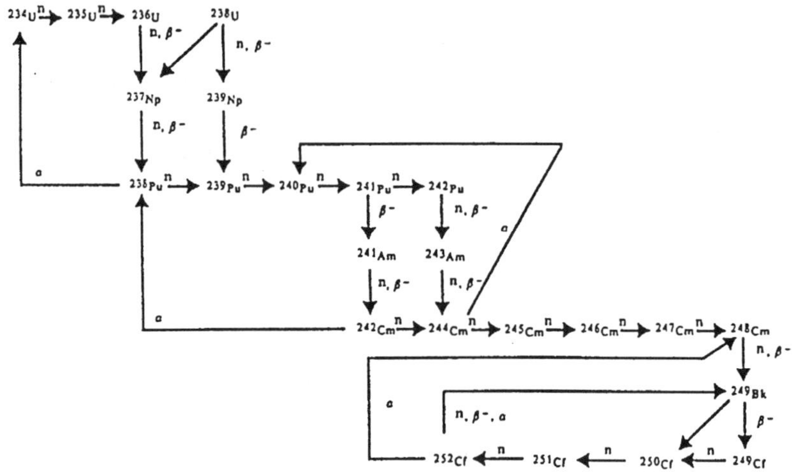

Fig. 1. Decay and neutron capture chains for actinides.

Since the minor actinides fission well in Liquid Fast Reactors (LFRs), it is possible to add small quantities of chemically separated actinides (Np, Am, Cm, Cf) to the fuel and to efficiently turn them into fission products.

In general, the long-lived fission products have small reaction cross sections for the processes of interest, i.e. (n, g), (n,2n), (n,p), and (n, a).

$$^{99}\text{Tc} + \text{n} \longrightarrow \,^{100}\text{Tc} \xrightarrow[16\,\text{s}]{\beta^-} \,^{100}\text{Ru (stable)} \qquad (1)$$

$$^{129}\text{I} + \text{n} \longrightarrow \,^{130}\text{I} \xrightarrow[12,4\,\text{h}]{\beta^-} \,^{130}\text{Xe (stable)} \qquad (2)$$

Further irradiation of Ru-100 and Xe-130 produces only stable isotopes.

For transmutation, the Tc-99 and I-129 would probably be placed in special target pins. An advantageous target form for technetium is certainly the metal, whereas for iodine a suitable form has still to be found. Iodine may react with potential cladding materials. Also, high pressure build-up in the pins may become a serious problem. Further extensive research on this matter is required.

4. Strategies for Nuclear Waste Management

Pursuing a closed nuclear fuel cycle through partitioning and transmutation appears to be an attractive strategy for future advanced waste management. It seems appropriate to distinguish between:

- Partitioning and more efficient element separation, respectively, on the one hand as the first practical priority; and
- Transmutation, on the other hand, as the second priority task.

Because the transmutation rate decreases with irradiation time, a 100 % transmutation in a single irradiation is impossible. Furthermore, if not very short-lived, the transmutation product will increase in amount with time and also transmute; usually an undesirable effect. Untransmuted material must thus be separated from transmutation products and recycled for further irradiation.

The less material that can be transmuted in a single irradiation, the more cycles are needed to destroy a given amount. Each time the material passes through one cycle, there will be some loss of untransmuted material to various process waste streams. These waste streams require appropriate treatment and subsequent ultimate disposal since their further recycling would become too complicated and costly. Because of the inevitable limit on the achievable combined recycling yield, reprocessing yields should be forced up as high as possible (> 99.5 %), otherwise the net gain of transmutation could vanish.

The merits of a more quantitative separation of Np, Pu and Am from the PUREX process high-activity raffinate waste stream has been investigated by the French CASTAING group [9]. Their findings are depicted in Fig. 2.

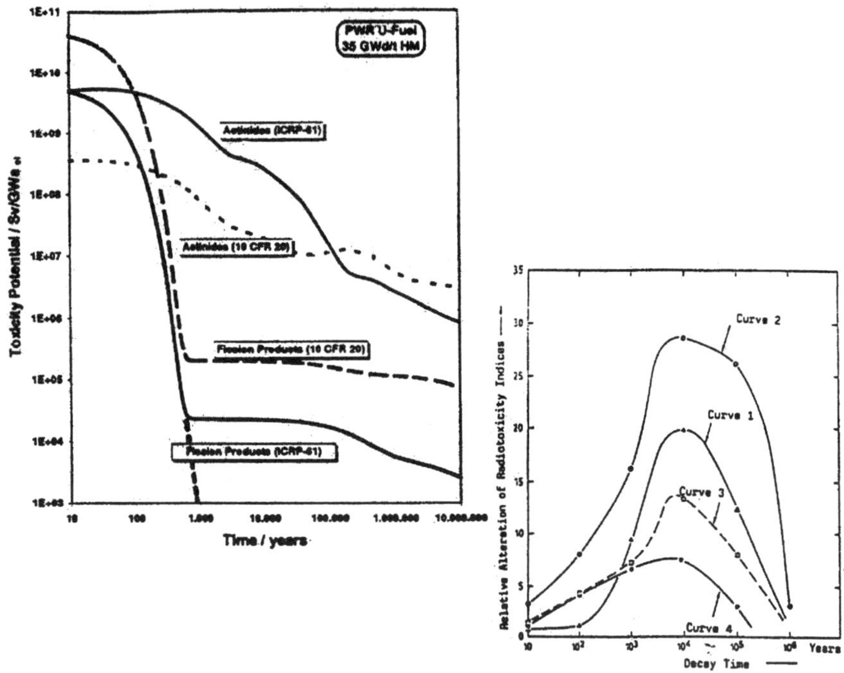

Fig. 2. Model calculations of radioactivity release from a repository, alteration of radiotoxicity indices as a function of separation factors based on ICRP recommendations.

Curve 1:	irradiation of fuel assuming ICRP recommendations No. 2 + 6
Curve 2:	irradiation of fuel assuming ICRP recommendation No. 30
Curve 3:	vitrified waste with 99 % Np and 98 % Pu removal, ICRP recommendation No. 30
Curve 4:	vitrified waste with 99 % removal of Np, Pu and Am, ICRP recommendation No. 30

A further reduction of factor 2 to 4 could be achieved if the removal of all relevant actinides were increased to 99%. However, one should not overlook the fact that 1 to 3% of the total transuranic elements present in the irradiated fuel are routed to waste streams other than the liquid HAW raffinate stream, e.g. medium activity waste streams arising during fuel fabrication and reprocessing not accessible for transmutation. Thus, the achievable overall radiotoxicity reduction is dictated by the lost actinide quantity spread over the mentioned waste streams.

A remedial measure is provided by a closed advanced, non-aqueous process cycle as shown in Figure 3, which means pyroprocessing instead of the presently favored aqueous fuel reprocessing [6, 10].

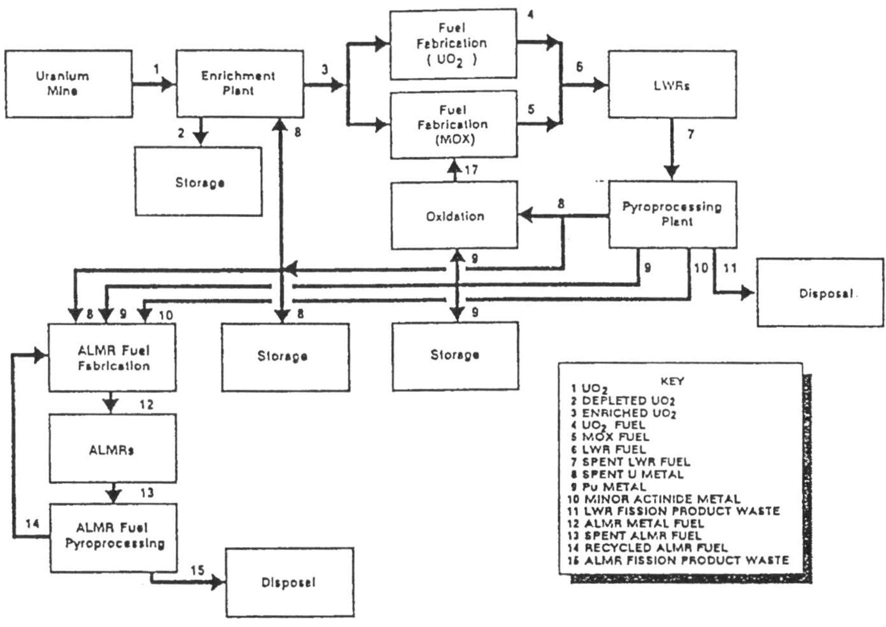

Fig. 3. Fuel and waste flow diagram based upon pyroprocessing.

To achieve a major actinide unit reduction factor of 10, it will be necessary to re-feed the initial load 8 times and to carry out eight cycles of irradiation, cooling, pyrometallurgical reprocessing and fuel refabrication.

Assuming further application of present day reprocessing technology, exhaustive partitioning and clean separation, respectively, of the actinides and long-lived fission products, this approach would represent an add-on step to existing technology. A more futuristic approach would be the development of a completely novel integrated chemical flow-sheet. Corresponding proposals have already been made but none of them has proven to be functional as yet.

5. Chemical Separation, Partitioning

Partitioning will be one of the key issues regarding the feasibility of the transmutation approach. It is important to have a separation process for radionuclides that can meet the required separation and decontamination factors without producing secondary radioactive wastes. Both aqueous-processing and pyro-processing are considered. The proposed process will be different depending on the transmutation concept and the considered transmutation strategies [11].

New, advanced chemical separation processes are needed to tackle the preset goal with regard to an efficient (99.9 %) transuranic element separation and special fission products from spent fuel.

5.1. NEPTUNIUM ISOLATION

The interaction of the PUREX process chemistry with neptunium redox reactions complicates the behavior of Np during this process. Neptunium can exist in three major oxidation states in an aqueous nitrate medium: Np^{4+}, NpO_2^+, NpO_2^{2+}. Fuel dissolution under oxidizing conditions would favor the higher oxidation states, preferably Np^{VI}, which is highly extractable in the HNO_3/TBP system, while the reductant added to ensure that all the plutonium is present in the extractable tetravalent state would then cause reduction to Np^V. In this five-valence state, Np is essentially inextractable. Np^{IV} is moderately extractable.

However, disproportionation of Np^V occurs, and traces of nitrous acid produced by nitric acid radiolysis will cause further redox reactions. The extent of all possible reactions is constrained by kinetic and equilibrium effects and is influenced by temperature and radiation levels.

The actual behavior in the PUREX flowsheet is such that part of the neptunium is extracted into the organic phase together with plutonium and uranium in the first extraction cycle. Depending on the prevailing conditions, the distribution between the aqueous raffinate and the organic phase may differ to a large extent. In any case, about equal parts of the neptunium follow the uranium and plutonium in the product streams.

Many chemical flowsheets for the recovery of neptunium reflect the efforts to lead this element into the highly active waste stream and dispose of it with the fission products. If a neptunium partitioning for the sake of later transmutation was to be strived for, selective neptunium isolation applying a suitable solvent extraction procedure might offer a solution. However, such a processing path still has to be developed.

An alternative approach for clean and quantitative neptunium isolation could be a selective back-extraction of Np^{IV} from the organic TBP/kerosene phase. This approach requires as a prerequisite, e.g., a quantitative routing of neptunium into the organic phase. Subsequently, a selective back-extraction of either Np^{IV} or Np^{VI} could be performed after the necessary valency adjustment via a suitable treatment has been successful. For a final purification of neptunium, an anion-ion exchange treatment is recommended.

Although several proposals have been put forward for complete recovery of neptunium from spent fuel solutions, none of them is completely satisfactory. More R&D efforts are therefore needed.

It should be mentioned that the methods of HPLC (high pressure liquid chromatography) and reverse column extraction chromatography have also been proposed to solve the problem of clean neptunium isolation. They have proved very useful for analytical purposes as well as for laboratory scale separations of neptunium from fission products and actinides. However, they are not qualified for application on an industrial scale.

5.2. MINOR ACTINIDES ISOLATION

Current concepts for high-efficiency separation of actinides call for improved plutonium recovery, coextraction of uranium and neptunium with subsequent partitioning by valence control, and extraction of americium, curium and californium from the HAW stream. There are a number of major problems to be solved before a technically feasible process is available.

For the recovery of Am, Cm and Cf from the waste stream, cation-exchange and extraction processes appear most promising. The major problem is to achieve a highly effective separation of actinides from lanthanides. An actinide/lanthanide fraction would probably have to be separated first from the other fission products and waste components and then the actinides would have to be recovered with a high yield and purity. Also, taking into account the fact that substantial additional waste streams would have to be managed without significantly increasing the overall waste quantity, it is obvious that the recovery of americium, curium and possibly californium will be the most difficult task in waste partitioning.

A series of extractants has been investigated in many laboratories with the aim of partitioning of actinides from both spent fuel and waste solutions. Details will not be given here, but rather a listing (see Table 1) of the various extractants for the purpose of an overview. A literature review has been published by Z. Kolarik "Separation of Actinides and Long-Lived Fission Products from High Level Radioactive Wastes" [11].

5.2.1 *Examples of flowsheets of solvent extraction processes.*

Numerous unit operations are available for the planning of partitioning processes. They deal with processes for the joint extraction of lanthanides(III) and transuranic elements(III) as well as with selective extraction of transplutonides(III) or lanthanides(III). Numerous flowsheets with many variations can be designed and consequently, numerous flowsheets have already been published. Only a few of them have been tested in hot experiments with real HAW solutions, many of them have only been tested in cold runs with simulated feed solutions, and some of them have not even been tested at all (see Figs. 4-6).

TABLE 1. Solvating extractants

<table>
<tr><td>

Monofunctional Phosphoryl Extractants
     Tri-n-butylphosphate, TBP
     Trialkylphosphine oxides
     Di-isodecyl phosphoric acid, DIDPA
     Di-isodoctyl methyl phosphonate, DIOMP

Bifunctional Diphosphoryl or Phosphoryl-Carbonyl Extractants
     Carbomoyl methyl phosphine oxide, CMPO
     Tetrapentyl methylene diphosphonate in diethylbenzene
     Dihexyl-N, N-diethylcarbamoyl methylphosphonate
     Dibutyl-N, N-diethylcarbamoyl methylphosphonate
     N-octyl(phenyl)N,N-diisobutylcarbamoyl methylphosphine oxide

Bifunctional Diamide Extractants
     1-malonic acid sym-dioctyldimethyldiamide
     2-oxalic acid sym-dibutyldimethyldiamide
     3-(3-oxapentyl)malonic acid sym-dibutyldimethyldiamide

Acidic Extractants
     Dibutyl phosphoric acid
     Di(2-ethylhexyl)phosphoric acid, HDEHP
     1-phenyl-3-methyl-4-acetyl-5-pyrazolone
     Thenoyltrifluoroacetone, TTA
     Neocupferron
     N-benzoylphenylhydroxylamine
     N-2,4-dichlorobenzoylphenylhydroxylamine
     b-isopropyltropolone
     1-hydroxy-2-naphthoic acid
     2-hydroxy-1-naphthoic acid
     3-hydroxy-2-naphtohic acid
     5,7-dichloro-8-hydroxyquinoline

</td></tr>
</table>

For illustrative purposes, a few flowsheet examples are depicted. Presumably, less complex flowsheets will be required in future partitioning research.

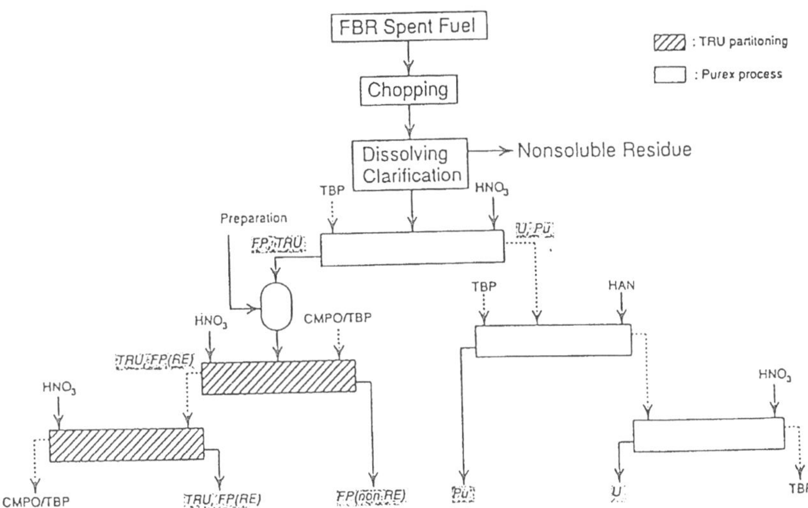

Fig. 4. Schematic block diagram of Japanese reprocessing experiment with FBR fuel performed at the PNC chemical processing facility [12].

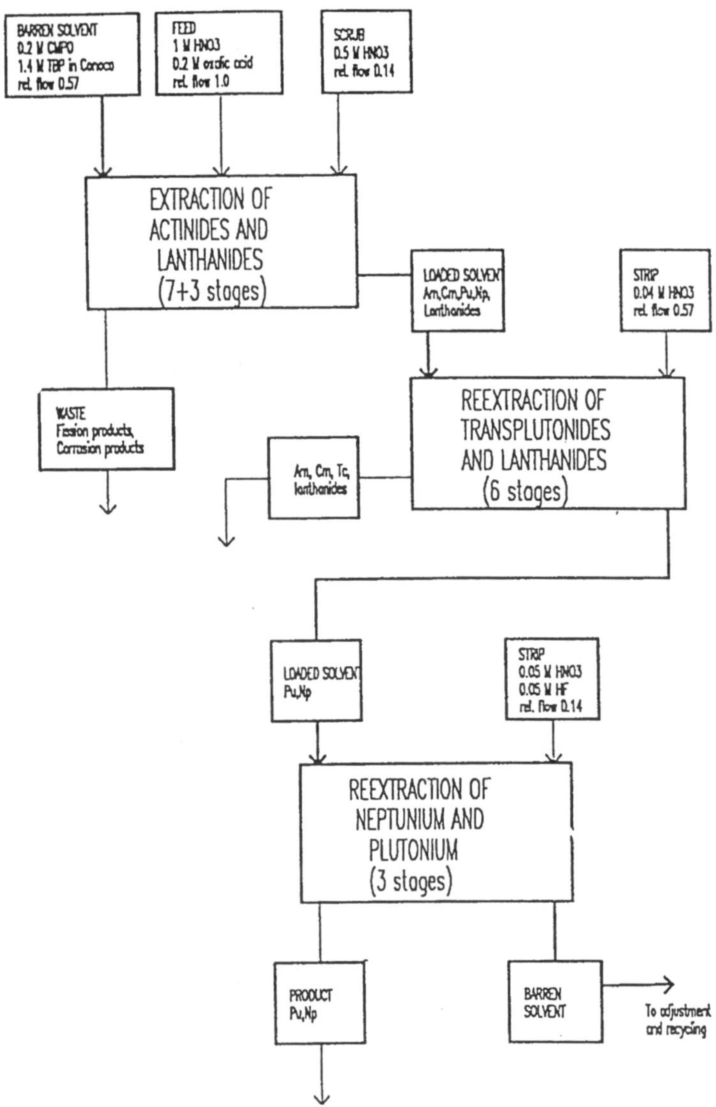

Fig. 5. HAW partitioning without transplutonide(III)-lanthanide(III) separation: Flowsheet of the TRUEX process (Argonne National Laboratory, Argonne and Rockwell, Hanford, USA) Conoco refers to a mixture of C_{12} - C_{14} n-paraffine, the feed is a HAW solution with oxalic acid added [13].

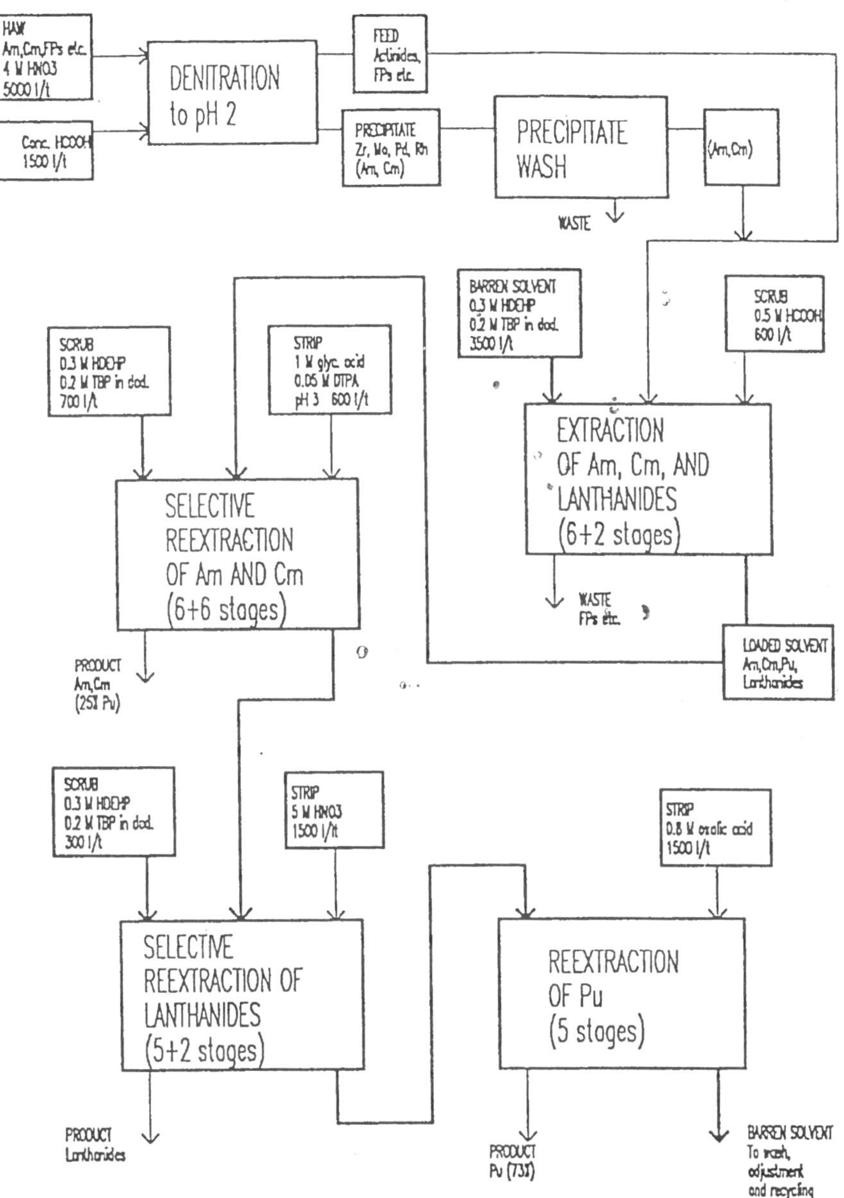

Fig. 6. HAW partitioning with transplutonide(III)/lanthanide(III) separation: Flowsheet of a process for the removal of actinides from HAW solutions (developed at EURATOM in Ispra, Italy). Dod. Refers to dodecane and glyc. acid refers to glycolic acid [14].

5.3. TECHNETIUM ISOLATION

It is widely recognized that technetium (Tc-99) represents a potential long-term hazard due to its long half-life and high chemical mobility in a geological environment. Therefore, a selective recovery in the course of fuel reprocessing is desirable for a follow-up transmutation.

Technetium is present in HAW solution as pertechnetic acid. The stable Tc(VII) oxidation state at 0.2 - 1.5 M HNO3 is extractable with TBP in dodecane. Thus, the majority of the Tc is found in the HAW raffinate. The remainder is extracted together with uranium and plutonium. However, one must note that a variable percentage of technetium, between 10 and 40 %, will be left in the dissolver insolubles, the so-called feed sludge [15].

During reactor irradiation technetium forms an intermetallic compound composed of Mo-Pd-Ru-Rh-Tc, which is more or less insoluble in nitric acid. Therefore, if one asks for complete Tc recovery, this insoluble portion must be isolated and treated in a suitable way to get it dissolved and combined with the bulk of technetium for further purification.

A selective extraction of TcO_4^- from the nitric acid HAW solution is possible using trioctylamine in CCl_4 as an extractant.

Another possibility for the recovery of technetium from aqueous solutions is offered by applying the electrolytic deposition of technetium as a metal or TcO_2, depending on the electrode materials and solution composition used. The method is hampered by the simultaneous deposition of the platinum metals Pd, Ru and Rh.

5.4. IODINE ISOLATION

Removal of radioiodine in the course of reprocessing could become complicated because of the numerous process streams in which iodine may appear and the variety of chemical forms it assumes. If iodine is allowed to remain in the solvent extraction feed, it reacts with solvent (TBP/kerosene) to form hard-to-remove compounds that eventually contaminate the whole system. It is thus important to remove as much of the iodine as possible before solvent contacting. Iodine may appear as I_2, HI and HIO, or organic iodines in off-gases or aqueous or organic phases, or as HIO_3 in concentrated nitric acid solutions [16].

The preferred procedure for removing iodine is to route it quantitatively into the off-gas during fuel dissolution with HNO_3. Practically all iodine from spent fuel will be released upon dissolution with the dissolver off-gas in the form of volatile elementary iodine, I_2. Special treatment involving sparging of the solution with NO_x-gas and adding of inactive iodine carrier is required to obtain better than 99 % iodine isolation. Problems may arise if iodine-129 is bound within an organic compound.

The released iodine may be fixed through absorption by aqueous sodium hydroxide. Another method uses 14 M HNO_3 containing 0.2 to 0.4 M $Hg(NO_3)_2$ to absorb all forms of iodine as HgI_2. Also sorption on silver-impregnated absorbents, such as zeolites, silica or alumina, are suitable for iodine fixation as AgI and AIO_3. The product is a dry solid, the chemisorbed iodine is highly insoluble in pure water, but it can be redissolved by forming soluble chlorine complexes in concentrated salt brine solutions according to the equations:

$$AgI + n\,Cl^- \leftrightarrow [AgI(Cl)_n]^n$$
$$AgI + n\,Cl^- \leftrightarrow [AgCl_n]^{n-1} + I^-$$

$$(3)$$

Iodine may be completely transformed into the dissociated I^- form by adding a cyanide reagent

$$[AgI(Cl)_n]^n + 2\,CN^- \quad [Ag(CN)_2]^- + n\,Cl + I^-$$

$$(4)$$

Finally, either elementary I_2 or a suitable chemical compound, e.g. PbI_2, may be isolated which can serve as a transmutation target material.

A great research effort is still needed before selective and quantitative iodine partitioning and also a suitable target preparation for subsequent transmutation reach maturity.

6. Pyroprocessing

Non-aqueous processing of spent reactor fuel has been considered to be a promising partitioning approach since the beginning of reprocessing technology. Pyroprocessing involves two different types of procedures:

- Metallurgy with oxidation/reduction reactions and thus slag formation and subsequent separation of slag from metal;
- Two-phase extraction of specific desirable or undesirable elements by contacting molten metal with a molten salt followed by separation of unreacted metal.

The first kind of process has been used for the reprocessing of metallic fuel from the US EBR-II reactor.

The driving force towards pyroprocessing is advanced molten salt reactor systems in which the materials undergoing transmutation in the high neutron flux would be continuously circulated through the transmuter blanket. It can be shown theoretically that homogeneous recycling is preferable for several reasons.

The transmuter blanket is continuously processed in order to achieve a real-time management of the transmuter volume contents. By doing so, transmuted elements can be removed as soon as they become stable and before they reabsorb significant numbers of neutrons to become radioactive again. The molten salt carrier medium is compatible with a highly versatile fluoride system. This system may be advantageous for various clean chemical partitioning schemes including simple distillation, differential aqueous extraction and ion exchange polishing of product streams.

Two ongoing projects, one in Japan and the other in the United States may illustrate the design of integrated partitioning concepts. The Japanese partitioning research project at the Central Research Institute of Electric Power Industry (CRIEPI) deals with dry processing for transuranium element separation. The process schematic is shown in Fig. 7.

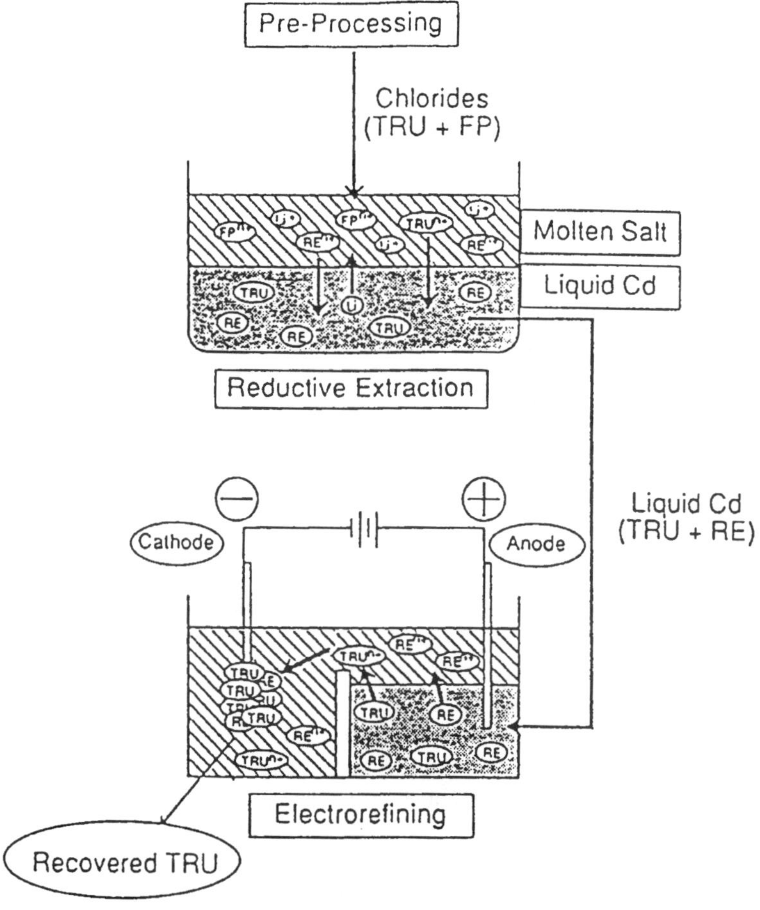

Fig. 7. Schematic illustration of a pyrometallurgical partitioning process [17].

The process is designed to remove 99 % of each actinide from a PUREX type high level radioactive waste raffinate. The envisaged goal is an actinide product that contains less than 10 % rare earths. The molten salt system consists of a KCl-LiCl eutectic salt mixture and liquid cadmium metal. A conceptual design study of a waste actinide burning fast reactor that can achieve criticality with pure transuranium element fuel is in progress.

A different reactor and fuel cycle concept is presently under investigation in the United States. The so-called Integrated Fast Reactor (IFR) concept involves a pool-type sodium-cooled reactor using a metal fuel and an on-site pyrometallurgical procedure for reprocessing and recycling of the fuel and blanket. A schematic drawing of the IFR fuel cycle is shown in Fig. 8.

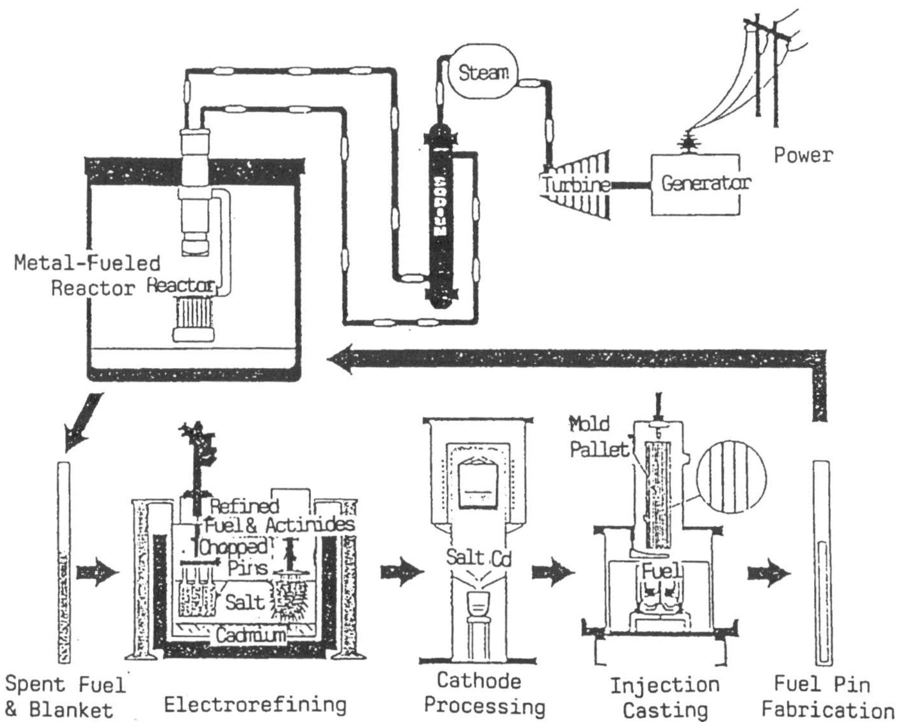

Fig. 8.   Schematic design of the IFR concept (Argonne National Laboratory) [18].

A key feature of the IFR concept is the pyrometallurgical reprocessing, which employs electrorefining at 500°C, using a cadmium anode and a molten salt electrolyte ($KCl-LiCl-U/PuCl_3$). Minor actinides will accompany the plutonium product stream and thus actinide recycling seems to be possible.

In conclusion, feasibility studies performed so far have revealed a great potential concerning the advanced integrated reactor and fuel cycle systems which allow partitioning and transmutation of actinides as well as some long-lived fission products. Nevertheless, the development and implementation of a partitioning/transmutation system would require a long-term research effort. There is no clear-cut view about the difficulties encountered on the chemistry part. Some experts are optimistic, others completely reject the possibility of a successful realization.

7. Conclusions

Partitioning and transmutation of transuranic elements (minor actinides) and some long-lived fission products (mainly Tc-99 and I-129) would help to reduce the quantity of these radionuclides to be disposed of in geological repositories. It also could reduce the levels of radiation exposure that might be received by the population in the distant future.

However, when all radiological aspects, both for the workers in the fuel processing plants and the general public, are considered in an overall assessment, together with all other relevant nuclear safety, technical and economic factors, this benefit might be outweighed by the additional risks and costs introduced. On the basis of current knowledge, a radioactive waste management strategy involving the partitioning and transmutation of actinides and some long-lived fission products is not justified in the context of an overall risk/cost/benefit analysis.

Partitioning and transmutation should only be introduced, if at all, as part of a long-term decision about using new nuclear power technology as a future energy source. With regard to this, R&D work dealing with basic chemical questions seems to be worthwhile and justified.

Insurance against the unknown is the most powerful argument for partitioning and transmutation. By removing long-lived isotopes from nuclear waste one will have an unchallengeable ability to predict future effects due to the simple fact that if radionuclides do not exist, they cannot cause any effects. The introduction of partitioning and transmutation will not eliminate the need for radioactive waste disposal.

References

1. Skalberg, J. O. Liljenzin: Partitioning and transmutation. A review of the current state of the art. SKB Technical Report 92-19, Stockholm, Sweden (1992)

2. PAGIS, Performance Assessment of Geological Isolation Systems for Radioactive Waste, Summary. Commission of the European Communities, Report EUR 11775 EN (1988)

3. A. G. Croff: Historical Perspective on Partitioning-Transmutation, in C.W. Forsberg et al., Historical Perspective, Economic Analysis and Regulatory Analysis of the Impacts of Waste Partitioning-Transmutation on the Disposal of Radioactive Wastes, Oak Ridge National Laboratory Report ORNL TM-11650, October 1990

4. OECD Nuclear Energy Agency: Proceedings of the Information Exchange Meeting on Actinide and Fission Product Separation and Transmutation, Mito City, Japan, 6 - 8 Nov. 1990, OECD Nuclear Energy Agency, Paris 1991

5. International Atomic Energy Agency: Evaluation of Actinide Partitioning and Transmutation (Final Report of a Coordinated Research Programme on Environmental Evaluation and Hazard Assessment of the Separation of Actinides from Nuclear Wastes Followed by Either Transmutation or Separate Disposal), Technical Reports Series No. 214, International Atomic Energy Agency, STI DOC 10 214, ISBN 92-0-125182-3, Vienna 1982

6. USDOE: Nuclear Waste Management with Actinide Conversion, Rockwell International, Report No. AI-DOE-13568, November 1989

7. E. D. Arthur: Summary, R&D Issues, Los Alamos Concept for Accelerator Transmutation of Waste and Fission Energy Production, Proceedings of the Specialist Meeting on Accelerator-Driven Transmutation Technology for Radwaste and other Applications, Saltsjobaden, Stockholm, Sweden 24 - 28 June 1991, Los Alamos National Laboratory Report LA-12205-C and Statens Karnbranslenamds Report SKN No. 54, November 1991

8. L. H. Baetsle: Impact of Fission Product Partitioning and Transmutation of Np-237, I-129 and Tc-99 on Waste Disposal Strategies, Proceedings of the Information Exchange Meeting on Actinide and Fission Product Separation and Transmutation, Mito City, Japan, 6 - 8 Nov. 1990, OECD Nuclear Energy Agency, p. 299, Paris 1991, also in (Belgian Nuclear Research Center-Belgium), Presented at the IAEA Advisory Group Meeting on Partitioning and Transmutation of Actinides and Selected Fission Products from HLW, Vienna, Austria, 21 - 24 October, 1991.

9. CASTAING-Report: Ministere du Redeploiment Industriel et du Commerce Exterieur. Conceil Superieur de la Surete Nucleaire, "Rapport du Groupe de Travail sur les Recherches et Developments en Materiere de Gestion des Dechets Radioactifs". CEA Paris (1984)

10. L. Koch: Formation and Recycling of Minor Actinides in Nuclear Power Stations. Handbook of the Physics and Chemistry of the Actinides, Vol. 4, Chapter 9 Elsevier Science Publishers (1986)

11. Z. Kolarik: Separation of Actinides and Long-Lived Fissio Products from High-Level Radioactive Wastes (A Review), Report KfK 4945, November 1991

12. M. Ozawa, S. Nemoto, Y. Kuno: Status of Actinide Partitioning Study in PNC, PNC-Japan, Presented at the OECD Nuclear Energy Agency Workshop on Partitioning of Actinides and Fission Products, Mito City and the Tokai Research Establishment 12 - 21, Nov. 1991

13. E. P. Horwitz, D. G. Kalina, H. Diamond, L. Kaplan, G. P. Vandegrift, R. A. Leonard,M. J. Steindler and W. W. Schulz, Proc. Int. Symp. Actinide Lanthanide Sepns., Honolulu, Hawai, Dec. 16 - 22, 1984. World Scientific, Singapore (1985), p. 43

14. P. Barbero, L. Cecille, F. Mannone, G. Tanet, S. Valkiers and H. Willers, Proc. 2nd Techn. Meeting Nucl. Transmutation of Actinides (Report EUR 6929), Ispra, Italy, Apr. 21 - 24, 1980, p. 211

15. W. T. Smith, J. W. Cobble, G. E. Boyd: Technetium. J. Am. Chem. Soc. 75 (1953) 5773 - 5777H. Brucher, E. Merz: Entsorgungsstrategien fur radioaktive Sonderabfalle. Report JUL-2099, ISSN-0366-0885, November 1986

16. H. Brucher, E. Merz: Entsorgungsstratgien fur radioaktive Sonderabfalle. Report JUL-2099, ISSN-0366-0885, November 1986.

17. M. Sakata, H. Miyarshiroo, T. Inoue: Basic Concept of Partitioning and Transmutation Research in CRIEPI and Denitration and Chlorination Technology for Pyrometallurgical Partitioning, Proceedings of the Information Exchange Meeting on Actinide and Fission Product Separation and Transmutation, Mito City, Japan 6 - 8 Nov. 1990, OECD Nuclear Energy Agency, p. 210, Paris 1991.

18. Y. I. Chang: Actinide Recycle Potential in the IFR, ANL-USA, Presented at the Symposium on Separation Technology and Transmutation Systems (STATS), Washington D.C., USA, 13 - 14 January 1992.

NUCLEAR CRITICALITY SAFETY ASPECTS OF THE UTILIZATION OF WEAPONS-GRADE PLUTONIUM IN MOX FUEL USING CURRENT BNFL TECHNOLOGY

L. M. FARRINGTON
P. E. BROOME
British Nuclear Fuels plc
Risley
Warrington
Cheshire, WAS 6AS
United Kingdom

Abstract

British Nuclear Fuels plc (BNFL) is currently commissioning the Sellafield MOX Plant (SMP), which is due to commence operation in 1998. This plant is capable of supplying bulk orders of both PWR and BWR fuels to a wide range of specifications. The design and construction of this plant followed a decision by BNFL in 1989 to enter the thermal MOX fuel market and to become a world leader in MOX fuel supply. The intention was to process spent oxide fuel in BNFL's Thermal Oxide Reprocessing Plant (Thorp) and return the reprocessed products to its customers in the form of MOX fuel. Production of MOX fuel for use in commercial power reactors is a safe and efficient method of recycling the plutonium recovered during reprocessing.

Since 1989, major steps have been taken under nuclear weapons disarmament treaties to reduce the number of nuclear weapons in service in Russia and the United States and to dismantle the nuclear warheads. This has led to a significant and increasing stockpile of weapons-grade nuclear materials which are surplus to current military requirements in those countries. There is strong international interest in finding the most appropriate method of managing these surplus materials and a range of options have been proposed. One of these is to convert weapons-grade plutonium in the form of "pits" (nuclear warhead components) into civilian nuclear MOX fuel and to burn the plutonium in commercial nuclear reactors.

This has the advantage of converting the warhead material into a form which is less proliferation sensitive. For this reason, BNFL has examined the possibility of using MOX technology, developed for peaceful civilian use, for the conversion of warhead

65

T. A. Parish et al. (eds.),
Safety Issues Associated with Plutonium Involvement in the Nuclear Fuel Cycle, 65–76.
© 1999 *Kluwer Academic Publishers. Printed in the Netherlands.*

material. It is anticipated that this conversion will take place in the country of origin of the weapons-grade plutonium.

This paper details the design features, operational controls and protection systems within the Sellafield MOX Plant's design which ensure criticality safety during operations with plutonium arising from spent reactor fuel (civilian Pu, civilian plutonium). An outline is given of modifications which could be adopted to render such a plant design safely subcritical when processing weapons-grade plutonium. Criticality calculational methods are discussed, in brief, with particular reference to validation of these methods for MOX fuel applications.

1. Introduction

There is currently strong international interest in finding the most appropriate method of managing surplus Russian and American weapons-grade nuclear materials arising from the dismantling of their nuclear warheads. A range of options are being examined including continued storage, immobilization by vitrification and the conversion of the weapons-grade plutonium into civilian nuclear fuel.

The latter option involves incorporation of the plutonium into mixed plutonium and uranium oxide (MOX) fuel that could be burned in existing, or new, civilian reactors. Research and operating experience has confirmed that many modern reactors can be operated, with no detrimental effects upon performance, using a fuel made from PuO_2 in a UO_2 matrix.

The weapons-grade plutonium burned in this way will become degraded such that it will meet the spent fuel standard. Spent reactor fuel is highly radioactive due to the presence of large quantities of beta and gamma emitting fission products - it, therefore, has a high proliferation resistance. In addition, because of the long and ever-increasing irradiation times in civilian reactors, plutonium in the spent fuel becomes "contaminated" with the isotopes Pu-238, Pu-240 and Pu-242 which makes it less suitable for nuclear weapons.

Consequently, the aims of this plutonium disposition scenario are two-fold:

- To make use of the weapons-grade plutonium, which is an important source of energy, by producing MOX reactor fuel and using it to produce electricity; and
- To degrade the plutonium, thus bringing it into an equivalent isotopic state to that of civilian plutonium, which is far less accessible and attractive for weapons manufacture.

British Nuclear Fuels ple is currently commissioning a new commercial-scale MOX fuel fabrication facility at Sellafield. This state-of-the-art plant, the Sellafield MOX Plant (SMP), has been designed to produce high quality BWR and PWR MOX fuel pellets, rods and assemblies using plutonium recovered from the reprocessing of spent uranium fuel.

Nuclear criticality safety has been a fundamental consideration at all stages of the development of the SMP design and process. A major challenge for the criticality assessor was to identify a design which, for feed plutonium having widely varying isotopic compositions:

1. Ensured a high level of inherent safety;
2. Allowed a high plant throughput; and
3. Covered the entire range of customer specifications.

The current design has been developed and licensed to handle civilian plutonium which has an expected minimum of 17w/o Pu-240/Pu. This adequately encompasses the current Thorp baseload and, therefore, meets BNFL customer requirements. It is our belief that any future plant using the same basic design as SMP, modified to suit its own specific customer requirements, can be operated to safely process weapons-grade plutonium.

2. SMP Design and Control Features

For descriptive purposes the process used by SMP can be broken down into 5 main stages:

1. Powder receipt and processing;
2. Pellet production;
3. Rod fabrication;
4. Fuel element assembly; and
5. Assembly inspection, washing and ultimate dispatch.

Full descriptions of the SMP plant and process can be obtained elsewhere [1, 2]. The design details provided here relate mainly to the powder receipt and processing areas since these are key in the criticality safety strategy. Within these areas are placed criticality protection systems which act to limit the maximum achievable plutonium enrichment level in the homogenized MOX powder product.

The main process stages are:

- Receipt of the PuO_2 and UO_2 powder feeds and any required additives. There is also the potential to receive off-specification MOX material for recycling back into the process; and
- Controlled dispensing of these feeds in order to achieve the desired plutonium enrichment of the MOX fuel.

Weighed quantities of powder are transferred to the process tower which mixes and homogenizes the feed materials. A schematic diagram is included as Fig. 1 showing the major vessels in this area, i.e. the homogenization mill, homogenization blender, conditioning mill feed hopper and conditioning mill.

Throughout SMP, the issue of dose uptake (which is greatly dependent on the Pu-240 content) has strongly influenced the design approach. All reasonable steps have been taken to reduce doses to the operators and this has resulted in the requirement for a largely automatic and remotely operated plant. Within SMP this requirement is met by a multi-functional, software-based, process control system known as the Integrated Automated System (IAS). This system meets the highest standards of engineering design and ensures that a wide range of fuel can be produced with guaranteed high quality

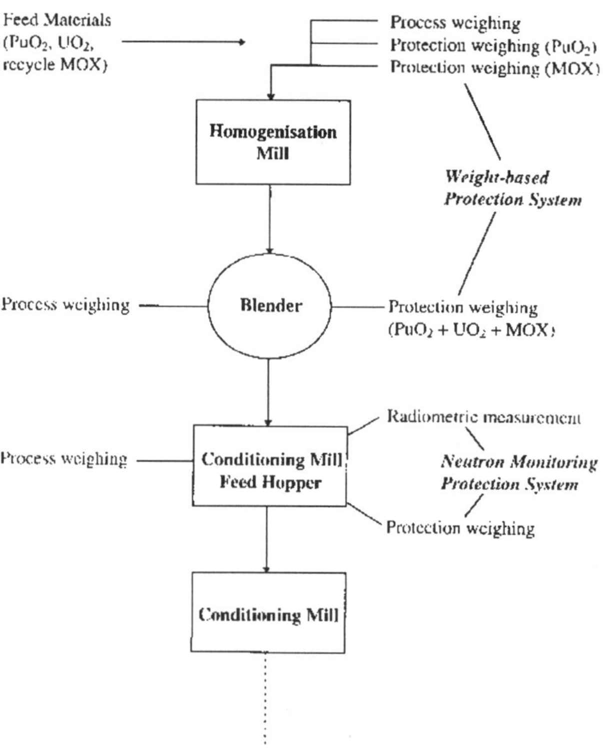

Fig. 1. Schematic of main process vessels in the SMP powder processing area.

The design intent was to have the capability to produce MOX fuel with plutonium enrichments up to a maximum of 10w/o Pu(fissile)/(Pu+U), using feed plutonium having a wide range of isotopic compositions. This was achievable with the IAS but, such a software-based control system is not an acceptable means of criticality safety control to the United Kingdom Regulators. Only limited reliability, in terms of the safety offered by such a system, would be justifiable. Moreover, due to the multi functional nature of the IAS, any failure or maloperation in its operation must be assumed to simultaneously affect a number of the control functions.

As a result of the limited reliability of the IAS, a series of extremely complex fault scenarios and physics related problems arose to challenge the production of a robust criticality safety case.

These problems were overcome by including a series of criticality protection systems which have the function of limiting the maximum plutonium enrichment of the MOX material in the event of multiple failure of the software system. These criticality protection systems are individually hard-wired into the process and operate totally independently of the IAS.

These systems are discussed below with particular reference to the Pu-240/Pu content of the feed plutonium. This is because, for criticality purposes, the Pu-240 content is the main area of interest:

- It influences the reactivity of any MOX produced; and
- It influences the range of possible enrichments that can be made in SMP.

BNFL have developed design principles for protection systems which are documented and fully endorsed by the Regulators. In the present application the protection system required was the highest category recommended by the design principles. This requires **two** independent protection methods with at least one measurement method completely different to that of the process control system.

This was achieved by including:

1. a protection system to restrict the maximum plutonium mass that can be added to the homogenization mill and to ensure that the correct batch size is produced. This system is referred to as the "weight-based" system; and
2. a neutron monitoring system to prevent the passing forward of MOX fuel with a plutonium enrichment greater than a predetermined limit. The characteristics of this instrument are such that it is necessary to set its trip level at a value somewhat greater than the design basis maximum enrichment of 10w/o Pu(fissile)/(Pu+U). It is indicated later (Section 2.2), that a trip level of 20w/o Pu(fissile)/(Pu+U) is appropriate. All SMP process areas have been designed to be safely subcritical when containing MOX material enriched to at least this enrichment.

2.1 WEIGHT-BASED PROTECTION SYSTEM

This protection system is designed to restrict the mass of fissile material which can be added to any mill batch. It is made up of two protection weighing systems, one on the feed to the homogenization mill and one on the blender:

- The masses of PuO_2 and recycle MOX which can be added to any mill batch are restricted to predetermined maximum values. These maximum masses are guaranteed by the protection system which weighs the feeds to the mill; and

- MOX material cannot leave the blender until a **minimum** weight of (PuO_2 + UO_2 + recycle MOX) has been confirmed by the weight protection measurement on the blender. This serves to ensure adequate "dilution" of the PuO_2 by UO_2.

The above mass limits were derived by assuming that the Pu-240 content of the feed PuO_2 could be as low as 17w/o Pu-240/Pu (with an upper bound of around 40w/o) and that the balance of the PuO_2 is Pu-239. The maximum achievable fissile content of the MOX material can be varied by appropriate choice of the protection system mass limits.

For any given mass of PuO_2 in the batch, the enrichment produced will be a function of the Pu-240/Pu content, i.e. as the Pu-240 content increases the enrichment of the MOX decreases. The overall behavior of this weight-based system as a function of Pu240/Pu content is shown in Fig. 2.

It can be seen from Fig. 2 that the design of the weight-based system is such that the maximum achievable MOX enrichment is slowly varying with quite large variations in Pu-240/Pu. At very low values of Pu-240/Pu, as may be appropriate for weapons-grade plutonium, achievable plutonium enrichments of the MOX do not exceed about 25%. It is also seen that it is possible to manufacture the design basis maximum MOX enrichment of 10w/o Pu(fissile)/(Pu+U) across the full range of Pu-240/Pu feedstock material.

2.2 NEUTRON MONITORING PROTECTION SYSTEM

This system is fitted to the conditioning mill feed hopper. Simplistically, it counts the neutrons emitted from spontaneous fission. The count rate is, thus, proportional to the mass of Pu-240 (and Pu-238) in the plutonium. The monitor is calibrated against a known Pu-240 content in a known mass of PuO_2.

To cover all possible feedstock the monitor calibration point is set for the "worst case" material. This will be the fuel with the **lowest** neutron count rate per unit mass, i.e. corresponding to the **highest** fissile plutonium content. Such material requires the least mass to reach the limiting MOX enrichment of 20w/o Pu(fissile)/(Pu+U). The calibration point is currently set at 17w/o Pu-240/Pu. The behavior of this neutron monitoring system as a function of Pu-240/Pu content is shown in Fig. 2.

It can be seen from Fig. 2 that the neutron monitoring system is very sensitive to changes in Pu-240/Pu. The design basis enrichment of 10w/o Pu(fissile)/(Pu+U) can only be manufactured with a limited, but acceptable, range of feedstock. This range is dependent on the chosen maximum enrichment calibration point. It is for this reason that an upper bound enrichment of 20w/o Pu(fissile)/(Pu+U) is adopted. Reducing this upper bound would reduce suitable feedstock to an unacceptable level.

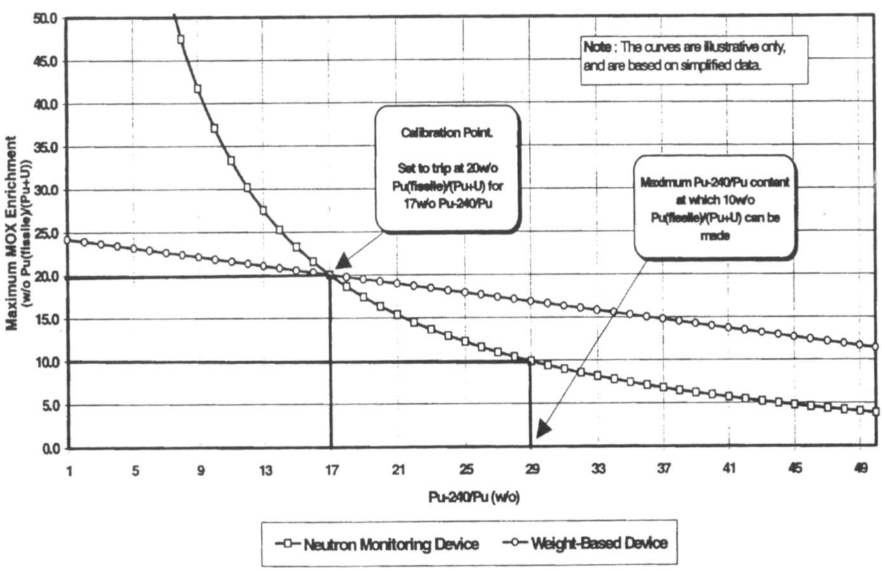

Fig. 2. Demonstration of maximum enrichments allowed by the protection devices as a function of PU-240/Pu content. Trip point is 17 w/o Pu-240/Pu.

72

The two protection systems acting in combination ensure that the probability of achieving an unsafe level of MOX enrichment is reduced to an acceptable value. It is necessary, however, to consider the single failure of each of the two systems. It is clear from Fig. 2 that such failures will **not** challenge the criticality safety of this plant design when it is processing the currently expected feedstock (i.e. >17w/o Pu-240/Pu), as the enrichment of 20w/o Pu(fissile)/(Pu+U) is never exceeded.

When considering weapons-grade plutonium, with its significantly lower Pu-240/Pu content, failure of the weight-based system, leading to total reliance on the neutron monitoring system, **would** present a challenge. This is because a low Pu-240/Pu will give a low count rate. The neutron system infers from the low count that the mass of PuO_2 is low and so will allow more to be added to the system. For the current configuration of the neutron monitoring system the result could be that highly enriched MOX material would be produced.

It should be noted that the plant design does offer a large degree of inherent safety. All single fault conditions are deterministically safely subcritical. A number of multiple fault conditions are similarly deterministically safe. There is also a series of mechanical interlocks which serve to prevent the occurrence of certain fault conditions.

Additional protection is provided by:

- Systems which restrict the masses of moderator added to the process; and
- Sampling operations.

The functionality of such protection is not directly, or adversely, affected by the isotopic composition of the feed plutonium and is, therefore, not discussed in detail here.

If the SMP design technology were to be utilized in the countries of origin of the weapons-grade plutonium, then modifications would need to be incorporated to ensure that nuclear criticality safety, as well as conventional safety, was not compromised. The degree of modification required will depend on a number of factors including specific customer requirements. Several design iterations have been considered by BNFL. These are discussed below.

3. Potential Strategy for Processing Weapons-grade Plutonium

Several options have been considered to give confidence that weapons-grade plutonium can be safely processed using the SMP design technology. It is not envisaged that these options will require any significant degree of re-design or engineering. The absolute degree will be a function of the requirements of the customer.

In order to optimize the required modifications to the design, BNFL have considered each process area in turn. Those areas where safety margins will be most significantly affected by a reduction in the Pu-240/Pu content of the feed plutonium have been identified. As expected with weapons-grade plutonium, the most sensitive area is the assembly washing process.

A number of potential change strategies were considered which, if implemented, would return the design to its original state in terms of nuclear and conventional safety.

These strategies are listed below, grouped according to the general function they perform.

The following options perform the function of off-setting the increase in reactivity caused by the use of weapons-grade plutonium by reducing the maximum achievable enrichment:

1. Recalibrate the enrichment protection systems;
2. Use operational experience to enable greater safety reliance on the IAS functions; and
3. Set the maximum design basis enrichment at a level appropriate to reactor requirements.

The options below serve to ensure that the feed material is no more onerous in criticality terms than feeding civilian plutonium:

1. Pre-mix the PuO_2 and UO_2 feed materials prior to the powder processing area; and
2. Blend civilian plutonium with weapons-grade plutonium prior to milling.

The following provide additional protection to compensate, in risk assessment terms, for the fact that the plutonium feed no longer has a guaranteed minimum Pu-240 content of 17w/o:

1. Supplement IAS functions with non-automated procedures, thus improving overall safety-related integrity. This is possible due to the lower doses inherent in weapons-grade plutonium;
2. Improve the overall integrity of the moderation protection system to ensure that conditions of excessive moderation are prevented; and
3. Improve the integrity of the sampling regime.

The options below serve to impact directly on the safety of the most limiting area of the entire process, i.e. that of fuel assembly washing. By eliminating the operation they remove the problem, but in so doing they restrict the scope and capability of the plant. It is acknowledged, however, that not all customers may require their assemblies to be washed:

1. Use an alternative to water immersion for the cleaning of fuel assemblies;
2. If water washing is the preferred alternative, then wash pins rather than complete assemblies;
3. Restrict the process to fabrication of fuel pellets or pins.

Some of the above proposals have different impacts on different parts of the process. Some also only offer benefit to specific process areas. For example, improved moderation protection will provide the necessary increase in criticality safety margin for the powder processing area, but will not benefit the fuel washing process. Conversely, an improved sampling regime will benefit the safety arguments related to fuel washing, with little beneficial impact on the powder processing area.

It should be noted that the above proposals do not represent an exhaustive list of possible design modifications. Additionally, the proposals have not been completely developed in terms of their design and process implications.

For these reasons, no single option is offered as the definitive solution. We are confident, however, that current BNFL technology and experience in designing and operating a similar plant, will enable customer-specific solutions to be found. This will be achieved by appropriate development of modification proposals based on the above examples.

All proposed design solutions will be assessed in terms of criticality safety by using well-established calculational techniques. The validity of these techniques when applied to the assessment of weapons-grade plutonium in MOX needs to be demonstrated. This issue is addressed in the following section.

4. Validation of Calculational Methods

Validation is the process of ensuring that the data, method of solution and the computer code or calculational route is adequate for the solution of a particular problem. This may be achieved by comparison with experimental data, standard analytical solutions or by comparison against other validated computer codes.

All criticality calculations completed in support of the safety case for SMP used the Monte Carlo neutronics code MONK7 [3].

In the context of criticality safety assessment, validation of complex computer codes, like MONK7, is achieved by direct comparison with high quality critical experiments. The current validation database for the MONK code and its associated nuclear database consists of critical experiments for uranium, plutonium and mixed systems over a range of moderation and reflection conditions. Each validation case provides a detailed representation of the critical configuration, including sensitivity analysis for all claimed experimental uncertainties.

MOX fuels present some interesting challenges to criticality codes, firstly because the nuclides involved in mixtures of uranium and plutonium can compete for the absorption of neutrons. Added to this, SMP has to deal with a range of moderation states from very dry powders to well-moderated arrays of pins. The validation cases that exist cover ranges of moderation states and ranges of enrichments, but there is inevitably some extrapolation or interpolation to be performed.

At the start of the criticality safety analysis of SMP some fourteen experiments were detailed in the validation database. There is an ongoing program of re-evaluation of these validation experiments and, each year, code user fees pay for additional experiments to be added to the database. BNFL have made a significant contribution above and beyond the normal fees so that priority could be given to MOX systems. A further 8 experiments have now been added to the database, some considering as many as 20 variations of the base experiment, to cover large ranges of moderation states (dry, intermediate and fully water saturated). In addition Pu/(Pu+U) ratios in the range 4 to

30w/o have been considered and Pu-240/Pu contents of the order 5 to 20w/o, thus providing data that is wholly appropriate to both civilian plutonium and weapons-grade plutonium.

The experimental data discussed above refers to that of mixed systems. There is also a significant amount of experimental data available for the separate MOX components, i.e. uranium and plutonium systems. Such data is valuable for MOX systems, particularly in fast and intermediate spectra where the main requirement is validation for Pu-239 and U-238.

Some of these experiments and the MOX experiments are currently being considered in a code comparison exercise to supplement the validation program. The Monte Carlo neutronics codes MONK7A and KENOVa (run as part of the SCALE 4.3 computer code suite) are being used in this exercise. The conclusion so far is that agreement between the codes is good, i.e. generally within 3 standard deviations of the calculated values of k-effective. Where greater differences are observed, MONK7A shows a tendency to slightly over-predict k-effective relative to KENOVa.

The conclusion of the MONK7 code **validation** exercise is that for dry, fast systems, similar to those in SMP, agreement between experiment and MONK is good with a slight tendency for over-prediction. For the intermediate energy range, the analysis again indicates an over-prediction. With regard to wet powders and MOX lattices, sensitivity profiles are very similar with a slight k-effective under-prediction of < 1% by MONK.

The approach adopted for SMP calculations has been to pessimistically assume that a reactivity under-prediction of 1% is appropriate across the entire range of applications. This is considered to allow for experimental uncertainty not covered in the sensitivity analysis, plus some contingency to cover small changes in sensitivity which may result from differences in physical form of materials, particularly between experiment and MOX powder applications.

5. Conclusions

This paper has outlined:

- the design features, operational controls and protection systems within the SMP design which ensure criticality safety during operations with plutonium arising from spent reactor fuel (civilian Pu,civilian plutonium);
- the various options which have been considered to give confidence that weapons grade plutonium can be safely processed using the SMP design technology;
- the criticality calculational methods which have been employed to demonstrate nuclear safety within SMP. Particular reference has been made to the validation of the preferred computer code, MONK7, for MOX fuel applications. The code is seen to predict the reactivity of MOX systems to a high degree of accuracy.

76

References

1. Farrington, L.M. and Munt, A. (1995). Novel Design Solutions for the Production of a Criticality Safety Case Supporting the Design of the Sellafield MOX Plant. *ICNC '95 The Fifth International Conference on Nuclear Criticality Safety, Albuquerque 1995,* Volume 2, 12.3-12.8.
2. Edge, J.A. and Walls, S.J. (1997). MOX Fuel Manufacture at Sellafield (to be issued as part of *"The 1997 ANS Nuclear Criticality and Safety Division Topical Meeting - Criticality Safety Challenges in the Next Decade", 7-11 September 1997, Lake Chelan, USA).*
3. ANSWERS/MONK(94)3 (1995). MONK - A Monte Carlo Program for Nuclear Criticality Safety Analyses. User guide for version 7A. Issue 2.

REDUCING THE PROLIFERATION RISK OF WEAPONS PLUTONIUM BY MIXING WITH REACTOR-GRADE PLUTONIUM

R. REIMERS
D. VON EHRENSTEIN
Kooperationsstelle Kerntechnik vod Energie an der Universität Bremen
Fachbereich Physik/Elektrotechnik
Universität Bremen, Postfach 330 440, 28334 Bremen
Germany.

Abstract

The following proposal is made to reduce significantly the proliferation incentive of excess weapons plutonium by mixing it with a comparable amount of separated reactor-grade plutonium. Degrading weapons plutonium by changing the isotopic composition will make the construction of nuclear weapons more difficult. It would not preclude any of the long-term disposal options under discussion (MOX or immobilization). This proposal would improve the present procedure for safeguarded storage because it could be implemented faster than any of the other options presently discussed. The immobilization option would benefit from this approach because it could then be applied to (nearly) reactor-grade and not to weapons-grade material. In particular, the simple reconstruction of dismantled weapons would be impossible because the degraded material could not be used in existing weapon designs.

1. The Plutonium Problem

The problem of disposition of excess weapons-grade plutonium and uranium has been investigated in studies by the US National Academy of Sciences [1], the German "Gesellschaft für Reaktorsicherheit" (GRS) together with the Russian ministry for nuclear energy (Minatom) [2], and the German American Council [3]. The commonly accepted standard for the envisaged proliferation resistance is the so-called "Spent Fuel Standard" (defined in [1]) which says in essence that any disposition measure must make

T. A. Parish et al. (eds.),
Safety Issues Associated with Plutonium Involvement in the Nuclear Fuel Cycle, 77–84.
© 1999 *Kluwer Academic Publishers. Printed in the Netherlands.*

the production of a nuclear weapon from this material as difficult as making it from (civilian reactor plutonium) spent fuel.

These studies have different assessments of whether or not the weapons plutonium has an economic value as nuclear fuel in the civilian fuel cycle. But regardless of the position taken on this issue, there is a general consensus that weapons plutonium poses a serious, proliferation problem, because rapid solutions to transform it into a less proliferation-prone state are not in sight. This aspect should have priority over all other considerations.

Presently, *safeguarded storage is* the only way to prevent misuse of the plutonium from the dismantled nuclear weapons, a temporary solution which is not regarded as very safe by the authors of [1]. They recommend to *"minimize the time during which the plutonium is stored in forms readily usable for nuclear weapons."*

The estimates for the amounts of weapons plutonium that will drop out of the nuclear weapons domain through 2003 (the scheduled year for the end of the START II implementation) are 100 t for Russia and about 50 t for the US [2, 3].[1] In order to attain the spent fuel standard, it was suggested that this material be used either as MOX fuel in light water reactors (LWR) or in direct disposal by immobilization together with highly radioactive waste.[2]

The prospects for the MOX option are doubtful, because it suffers from an insufficient civilian infrastructure. There are neither enough MOX fuel production facilities, nor enough LWRs with MOX license permission. Even if this bottleneck was removed, the weapon MOX-fuel would exhibit a high proliferation risk unless it had reached its scheduled burn-up. Furthermore, the implementation of this process will require at least 20-30 years.

Direct immobilization with high level radioactive waste will also need about 20 years, but it does not require introduction of weapons plutonium into the civilian fuel cycle. It has only one serious drawback. The spent fuel standard is achieved only for a limited time. The relevant isotopes in the high-level waste have half lives of around 30 years. After some 300 years, only one thousandth of these isotopes will remain, hence providing only a weak protection against any misuse. Then, the weapons-grade plutonium in the immobilization logs will be rather easily accessible.

2. The Mixing Proposal and Its Impacts on Proliferation Resistance

It is proposed here that the excess weapons plutonium be mixed with an approximately equal amount of separated reactor-grade plutonium from civil reprocessing.[3]

---

[1] It should be noted that the START Treaties impose a limit only on delivery vehicles not on warheads. The excess weapons plutonium problem is an inevitable consequence of this. At the time of writing this article, the US has already declared 38.2 t of weapons plutonium as being in excess.

[2] The US has decided to investigate both the MOX-LWR option and immobilization option to stay on the safe side in case either option should turn out to be impracticable [4].

[3] Similar proposals have been made previously by De Volpi in the late seventies [5].

The resulting mixture can then be classified as reactor-grade plutonium (of lower burnup). Construction of nuclear weapons is not impossible with this material, but the incentive to do so is greatly reduced. The following stages of the disposition process would benefit:

- The actual procedure for safeguarded storage: The gap in the proliferation resistance between the measures proposed for final disposition of the weapons plutonium and the actual procedure of safeguarded storage would be closed because mixing it could be implemented much earlier than any of the final disposition options mentioned above.

- The immobilization option (if chosen) would be applied to (nearly) reactor-grade plutonium instead of weapons-grade plutonium. The short-term recovery of such degraded plutonium is much less attractive because it cannot be reused in existing weapon designs. In the long term, the construction of new weapon types would be made more difficult. This fact will be of particular importance after the fission products that provide a passive radiation barrier against weapons use have decayed. A major weakness of the Spent Fuel Standard can be removed following this approach, because, according to its present formulation, it makes no provisions for the change of the isotopic composition of the plutonium.

Furthermore, the fabrication of MOX fuel would be facilitated because the system for preventing criticality accidents in MOX fuel plants uses the spontaneous fission neutrons from Pu-240 as an indicator [6].

This measure will provide enhanced protection against proliferation with respect to the following aspects:

1. *The suitability of the degraded plutonium for construction of nuclear weapon is greatly reduced: In particular the simple reconstruction of nuclear weapons of tested design is made impossible.* The most significant obstacle introduced following this approach is the higher neutron background. The risk of premature ignition is raised which makes the prediction of the yield uncertain and in effect may lead to a very low yield. This effect can only be compensated for by two measures: By improved performance of the igniting conventional charge or by boosting with tritium. In the first case, the requirements of the spherical symmetry for its implosion are higher and the compaction velocity must be roughly doubled [7, 8, 9]. In case of boosting, there will be the difficulty of obtaining sufficient amounts (some grams) of tritium.

---

[4]Provided that the research in laser separation of plutonium isotopes is ended. This is the only technology with the potential for plutonium re-separation. A laboratory operation has already been achieved in the US, but it is still years away from practical applicability. Research is only performed in the US, France, and South Africa [11].

80

Approximately twice as much reactor-grade plutonium than weapons-grade plutonium is required in comparable weapon assemblies due to the higher critical mass. This makes diversion more difficult because a larger amount of plutonium would have to be diverted. Together with the increased heat generation, such diversion would set upper and lower limits for the size of nuclear weapons produced from the diverted material. The IAEA calculates a requirement of 8 kg for reactor-grade plutonium whose diversion must not be undetected [10]. In [7] it is estimated, that a small nuclear weapon of reactor-grade plutonium (< 10 kg plutonium) does not need active cooling measures. Furthermore, the heat production limits the compactness of reactor-grade plutonium weapons because it is necessary to dissipate the heat.

2. *Isotopic degradation is the only irreversible degradation method.*[4] This distinction makes it superior to any method of spiking with strongly radioactive elements which can be reversed, in principle, by chemical means. This is the decisive advantage over chemical degradation and safeguarded storage. It is, therefore, a passive measure; the increase of the protection against proliferation depends on one single and irreversible step.

3. *Degradation of weapons-grade plutonium to reactor-grade plutonium increases the difficulty of processing and handling the fissile material.* Table 1 shows how neutron radiation and heat production can be raised in this fashion. Compared to ordinary weapons-grade plutonium, neutron radiation is increased by a factor of three and the heat production by a factor of three to five, depending on the burnup of the reactor-grade plutonium (and/or the percentage of added reactor plutonium).

TABLE 1. Composition, spontaneous fission neutrons, and heat production of weapons-grade plutonium, typical reactor-grade plutonium (from LWRs) and mixtures of weapons-grade plutonium with reactor-grade plutonium [12, 13]

	Composition of Pu arrays [in %]					Neutrons	Heat
	Pu-238	**Pu-239**	Pu-240	Pu-241	Pu-242	$[\frac{1}{\text{sec} \cdot \text{kg}}]$	$[\frac{W}{\text{kg}}]$
Weapongrade	–	**93.0**	6.5	0.5	–	104,000	2.23
33 GWd/t	1.3	**56.6**	23.2	13.9	4.7	495,000	10.5
45 GWd/t	3.1	**54.2**	19.4	15.5	7.8	548,000	20.4
Mixture 50/50 with 33 GWd/t Pu	0.7	**74.8**	14.9	7.2	2.4	303,000	6.67
Mixture 50/50 with 45 GWd/t Pu	1.6	**73.6**	13.0	8.0	3.9	328,000	11.6

According to [7], radiation doses are increased by a factor of 7.6 (in comparison to weapons-grade plutonium) when reactor-grade plutonium is processed without additional protection. The main contribution to the increased radioactivity is due to short-range X- and γ-radiation. The main contribution to this radiation is Am-241 which is a decay product of Pu-241. The intensity of this radiation increases over the years, as more and more Pu-241 decays into Am-241. A smaller contribution to the radiation dose has its origin in the increased number of neutrons from spontaneous fission of even-numbered isotopes. Comparatively simple shielding methods are sufficient for protection, and no additional shielding for material handling is necessary if an exposure 1 Sv is acceptable. Acute clinical radiation sickness symptoms may not necessarily be observed in this dose range.

The principal heat source in reactor-grade plutonium is Pu-238. Calculations [14] have shown that a 6 kg-sphere of metallic reactor-grade plutonium (33 GWd/t) has a temperature that is approximately 100 °C higher than ambient temperature. When plutonium is treated like a black body radiator, similar results (surface temperatures of 120-200 °C) are expected. These temperatures can be reduced by convection and other active cooling measures. The use of special handling tools is necessary. Processing large amounts of (nearly) reactor-grade plutonium requires inert gases.

4. *Mixing weapons-grade with reactor-grade plutonium would enhance safeguards verification because external radiation (neutron radiation by a factor of three) and heat production (by a factor of three to five) are increased.* Fetter *et al.* [15] have calculated the range of the neutron radiation from 4 kg of weapons-grade plutonium and from 12 kg highly enriched uranium in warheads shielded with tamper material typically used in nuclear weapons. They found a detectability up to a range of about 15-50 m with hand-held passive neutron detectors. This range would be extended to about 25-85 m if the plutonium mix was used instead[5]. An isotope occurring mainly in reactor-grade plutonium and emitting strong γ-radiation is americium-241 (decay product of plutonium-241), which can be detected by *using relatively simple portable* γ-detectors, although definite quantitative identification requires more sophisticated equipment.

The additional heat generated when Pu-238 is present can be detected with suitable equipment. Eight kilograms of isotopically degraded weapons-grade plutonium produces 50-90 W, which is comparable to a light bulb. Heat detectors should be capable of identifying the resulting "hot spot" in any assembly.

---

[5]This is a very   rough estimate basing only on the number of spontaneous fission neutrons occurring in mixed plutonium.

82

3. Perspectives for Short-term Realization

The separated reactor-grade plutonium needed for mixing has to be taken from the civilian sector. A necessary condition is an existing and functioning reprocessing facility. Presently, there are only two such facilities in the world which produce enough reactor-grade plutonium for admixture:

The French plants in La Hague and the British plant in Sellafield[6]. Since spent fuel from Germany (largest customer), Belgium, Switzerland, the Netherlands, and Japan is reprocessed at these facilities, the mixing option requires (besides the consent of the US and Russia) the consent of all these countries (plus France and Great Britain), provided their reprocessed fuel is needed. The World Information Service on Energy in Paris has published figures about the amount of reprocessed fuel in La Hague based on figures from COGEMA[7] [16, 17]. The estimate is that 57.3 t of reactor-grade plutonium has been separated as of June 30, 1994, which is compatible with approximately 1% of the reactor-grade plutonium in the reprocessed LWR fuel.

The operational records of Sellafield and La Hague show poor performance. There were great difficulties in Sellafield to reach the nominal capacity after the plant "THORP" (the only plant that can process LWR fuel) became operational in 1993. It had to be shut down most of the time. These technical difficulties seem to have been overcome the autumn of 1995. Since then, Sellafield has exhibited satisfactory performance. La Hague had to go through the same difficulties: During the seventies and eighties, the UP 2 plant did not perform at better than 20% of its nominal capacity (then 400 t/yr). The UP 3 plant was built in the late eighties, while at the same time the capacity of UP 2 was doubled (from 400 t to 800 t annually). Initial difficulties seem to have been overcome. Approximately 1280 t of spent fuel were reprocessed between 6/1/94 and 6/1/95, corresponding to roughly 11 t of separated reactor-grade plutonium. Assuming a similar performance for Sellafield yields an estimated 17 t of separated reactor-grade plutonium annually. Using this assumption until the year 2000, an estimated 150 t of reactor-grade plutonium from LWRs will have been separated in La Hague and Sellafield (THORP plant). An additional 25 t of separated reactor-grade plutonium could also be contributed by the Russian RT-1 plant near Chelyabinsk by the year 2000 [18]. Due to the rapid introduction of MOX fuel, particularly in France and Germany, a considerable portion of the separated reactor plutonium is no longer available for mixing. Currently about 100 t of separated reactor plutonium have been used for fueling reactors[8]. The mixing procedure can, therefore, proceed only slightly

---

[6]A third reprocessing plant for LWR fuel with an annual capacity of 800 t is scheduled to become operational in 2002 in Japan. Since the Japanese spent fuel policy is unclear and the reaching of the nominal capacity is likely to be even further in the future, we do not make allowances for this capacity here.

[7]COGEMA is the acronym for the national enterprise in charge of reprocessing in France.

[8]Including the French breeders Phenix and Super Phénix.

faster than the separation of reactor plutonium from spent fuel, that is at a rate of some 20 t annually. This rate, however, would mean that mixing could be completed in 2004 if begun in 1998. No new reprocessing plants would be needed.

Mixing could occur at MOX fuel element production plants. Full operation at such plants would not be needed for the proposed plutonium processing, since only a few components would have to be used. Reactor-grade plutonium is usually in the form of plutonium oxide (PuO_2), whereas weapons plutonium is in metallic form. Hence, the first step has to be the production of plutonium oxide from the weapons plutonium which, in principle, poses no technical difficulties. This operation can also be performed most conveniently in MOX fuel plants because they are designed to handle plutonium, thus making additional plutonium transport unnecessary. Only the initial stages of the MOX fuel factories would be needed since they were designed to produce MOX fuel of a well-defined composition. Thus, the technology can be taken directly from the existing infrastructure of MOX fuel production.

4. Final Remarks

This proposal has been conceived to avoid the MOX option, although it does not preclude any of the presently discussed disposal options. It is also designed to allay fears that the immobilization option might be chosen only, because it does not affect the weapons-grade plutonium quality. Such fears have surfaced particularly in Russia with respect to the US posture being in favor of immobilization rather than "MOXing" [19]. Possible objections against using plutonium from the presently operating civilian reprocessing plants should be seen as secondary in view of the necessity for rapid measures against the proliferation risks of excess weapons plutonium.

References

1. Committee on International Security and Arms Control (1994) Management and disposition of excess weapons plutonium. Technical report, National Academy of Sciences, Washington.
2. GRS, Minatom, and SIEMENS (1994) Technische Studie über die Produktion von Uran-Plutonium-Brennstoff aus waffengrädigem Plutonium und über die Moglichkeiten seines Einsatzes in der Kernenergiewirtschaft. Technical report, Gesellschaft für Reaktorsicherheit and Minatom, Hanau and Moscow.
3. Panofsky, W. K. H. and Soergel, V. (1995) U.S.-German cooperation in the elimination of excess weapons plutonium. Technical report, German American Council and U.S. National Academy of Sciences, Washington D.C..
4. Moore, M. (March/April 1997) Plutonium: The disposal decision. *Bulletin of the Atomic* Scientists, 53(3), 40ff..
5. De Volpi, A. (1979) *Proliferation Plutonium and Policy.* Pergamon Press, New York.

6. Farrington, L. M. and Broome, P. E. (1997) Nuclear criticality safety aspects of the utilization of weapons-grade plutonium in MOX fuel using current BNFL technology. In *Proceedings of the NATO Advanced Research Workshop on Safety Issues Associated with Plutonium Involvement in the Nuclear Fuel Cycle (Volga '97)*. Moscow, September 2-6.

7. Kankeleit, E., Küppers, C. and Imkeller, U. (1989) Bericht zur Waffentauglichkeit von Reaktorplutonium. IANUS-Working Paper 1/1989 (Extended Report for the expert hearing "Dangers of Spread of Nuclear Weapons" by the Hessian Landtag on 15.6.84.), Darmstadt. An English version is available. Translation prepared by the translation service of the University of Berkeley, California.

8. Mark, J. C. (1993) Explosive properties of reactor-grade plutonium. *Science & Global Security*, 4(1), 111.

9. Barleon, L., von Ehrenstein, D., Hahn, L. and Kankeleit, E. (1992) Zivile Nutzung der Kernenergie und ihre Proliferationsrisiken. In C. Eisenbart and D. von Ehrenstein, editors, *Nichtverbreitung von Nuklearwaffen - Krise eines & Konzepts*, Reihe A der Texte und Materialien der FEST, pages 257-302. Forschungsstätte der evangelischen Studiengemeinschaft, Heidelberg, 2nd edition.

10. IAEA-Staff (1993) Against the spread of nuclear weapons: IAEA safeguards in the 1990s. Brochure published by IAEA's Division of Public Information, Vienna, December.

11. Mohrhauer, H. (August/September 1995) Entwicklung bei der Uran-Anreicherung. *Atomwirtschaft-Atomtechnik*, 40(8/9), 537-541.

12. Taube, M. (1974) *Plutonium*. Verlag Chemie, Weinheim/Bergstr..

13. Albright, D., Berkhout, F. and Walker, W. (1997) *Plutonium and Highly Enriched Uranium 1996*. Oxford University Press, Oxford, Stockholm.

14. Nelson, W.E. (1977) *The Homemade Nuclear Bomb Syndrome*. PhD thesis, University of Missouri, Columbia. Cited according to [7].

15. Fetter, S., Prilutskii, F. and Rodionov, S. N. (1989) Passive detection of nuclear warheads. In J. Altmann and J. Rotblat, editors, *Verification of Arms Reductions*, page 48. Springer Verlag, Berlin, Heidelberg, New York.

16. Homberg, F., Pavageau, M. and Schneider, M. (December 1994) COGEMA-La Hague: Les Techniques de Production de Dechéts. Rapport réalise pour Greenpeace, World Information Service on Energy (WISE), Paris.

17. Schneider, M. (1997) Die deutsch-französische Atomfreundschaft. In W. Liebert and F. Schmithals, editors, *Tschernobyl und kein Ende 4*, pages 197-208, Münster. agenda Verlag.

18. Gelfort, E. and Krüger, F. W. (April 1997) Wiederaufarbeitung von Kernbrennstoffen in Russland. *Atomwirtschaft - Atomtechnik*, 42(4), 255.

19. Miller, M. and von Hippel, F. (July/August 1997) Let's reprocess the MOX plan. *Bulletin of the Atomic Scientists*, 53(5), 15-17.

COGEMA'S CONTRIBUTION TO THE RECYCLING OF MILITARY PLUTONIUM

C. de TURENNE
COGEMA
2, rue Paul-Dautier
78141 Velizy Cedex
France

1. Introduction

Signature of the START I and II treaties in 1991 and 1993 inaugurated a new era, marking a radical improvement in international security. Since then, and for the first time in history, nuclear disarmament has become a reality. This process includes the safe dismantling of thousands of warheads containing military grade nuclear materials consisiting of both uranium and plutonium.

In the case of highly enriched uranium (HEU), concrete actions have already been undertaken to recover potential energy contained through recycle as fuel for civilian reactors. This meeting is devoted to the technical challenges of safe and efficient utilization of plutonium in the fuel cycles of nuclear power plants, in relation with nuclear disarmament. COGEMA is here to present its experience in the use of civilian plutonium.

In our opinion, the technical feasibility of fabrication and irradiation of weapons grade plutonium fuel is for the main part demonstrated already at an industrial scale from civilian plutonium fuel experience.

COGEMA Group is ready to pursue with its partners, in cooperation with the governments of the US and Russia, Pu disposition in LWRs and in order to take advantage of the European MOX experience in their respective programs.

T. A. Parish et al. (eds.),
Safety Issues Associated with Plutonium Involvement in the Nuclear Fuel Cycle, 85–89.
© *1999 Kluwer Academic Publishers. Printed in the Netherlands.*

2. European MOX Experience

Plutonium recycling in the form of MOX fuel is now a mature industry in Europe, with successful operational experience at large-scale fabrication plants and successful power production from MOX assemblies in reactors.

2.1. FUEL IRRADIATION EXPERIENCE

In Europe today, up to 23 reactors are already loaded with MOX fuel, and 25 will be loaded by the end of this year with up to 50 to be loaded in the near future.

Today in France, 11 reactors (among 16 that are licensed) are loaded and 14 will be loaded by the end of this year, with up to 28 by 2005. In Germany, 7 reactors are using MOX fuel, (12 are licensed) and a total of 13 to 17 will be using MOX by 2005. In Switzerland, 4 reactors are licensed, 2 are currently loaded and one has just received its first MOX reload. In Belgium, two reactors have been licensed and have been loaded since 1995. At the end of 1997, about 700 t of MOX fuel will have been loaded in commercial light water reactors without any incident related to the nature of MOX fuel. In France, the mean burn-up of MOX fuel is 37.5 GWd/t In addition, four assemblies were loaded for a fourth cycle in 1994 and were unloaded with a confirmed burnup of 44.5 GWd/t.

2.2. MOX FUEL BEHAVIOR: COMPARISON WITH UO2

The Components and structure of MOX and UO2 assemblies are the same, with full compatibility in a reactor. The safety level of reactors loaded with 30% MOX fuel is preserved by adding control rod assemblies and by increasing the boron concentration in the tanks containing boric acid and water. Examinations in hot cells have shown that dimensional evolution and corrosion of MOX fuel rods were similar to those of UO2 fuel rods. Fission gas release is higher essentially due to the higher power delivered by MOX rods. This phenomenon was taken into account early in the conceptual design of the rods. As for pellet-cladding interaction, power ramp experiments have shown that MOX fuel behavior is more favorable than UO2 fuel.

2.3. MOX FUEL FABRICATION

COGEMA Group is currently operating two MOX fuel plants, one at Cadarache and a second called MELOX, located on the Marcoule site.

The capacity of the **Cadarache plant** can reach 40 t of MOX fuel per year of which 10 t is LMR fuels. This plant has already processed more than 30 tHM of plutonium, mainly for the production of about 100 t of FBR fuel. This experience has demonstrated a flawless industrial performance in plutonium recycling at Cadarache. By the end of 1997, Cadarache will have produced about 130 t of MOX fuel for LWRs.

The MELOX plant, brought on line in 1995, is the world's most efficient and modern MOX fabrication plant, truly bringing MOX fabrication to industrial maturity. Its new and automated design enables the plant to produce assemblies at a rate well above one assembly (500 kg) per day. Because of the good performance of this first industrial scale plant, COGEMA has decided to improve the plant with flexible line which will have the ability to produce different MOX fuel with different designs, and in particular, fuel assemblies for BWRs. For that, the capacity of the MELOX plant will reach about 250 t/y around year 2000.

COGEMA is also a partner of BelgoNucleaire (BN) through the commercial association COMMOX. BN operates a plant in Dessel, Belgium. This plant has a capacity of about 40 t/y and will reach a cumulative production since 1986 of around 350 t of MOX fuel at the end of this year.

All three plants currently use MIMAS process.

After more than thirty years of experience in MOX technology, the European industry can offer a mature fabrication technology and can adapt its operation for material of military origin, without introducing difficulties.

2.4 ADAPTATION TO WEAPONS GRADE PLUTONIUM

Even if the raw material is not absolutely similar, the civilian industry has enough experience to confidently expect that the disposition process can go forward without major obstacles.

Low content in isotopes Pu-238 and Pu-241, would even make the process easier in terms of thermal power, since it is 7 to 10 times lower in weapons-grade rather than civilian plutonium, and in terms of health physics and safety, because α-,γ - and neutrons emissions are also notably lower.

Beyond the conversion of plutonium metal to plutonium dioxide, the only constraint would come from criticality, leading to the need for some specific adaptation of the civilian MOX fabrication technology, for example smaller size equipment in the front part of the facility.

Criticality constraints have been taken into account while designing a modified process for civilian plutonium fuel. Introduction of military plutonium will simply mean in taking into account new parameters for the plant, but this is not expected to be hard to achieve by modifying the current design.

Years of experience in the fields of MOX fuel irradiation and of MOX fuel fabrication have given the European nuclear industry competence that can now become available for the WG-Pu military disposition programs in the US and Russia. COGEMA has already taken steps in that direction in both countries.

3. Russian Situation

3.1. AIDA MOX PROGRAM AND GRS/SIEMENS/MINATOM COOPERATION

In 1992, the French and German governments independently signed agreements with Russia on collaboration aimed toward peaceful utilization of ex-weapons material. The French program is called AIDA MOX. More precisely, this program aims at studying the feasibility of all steps needed for the Russian WG-Pu recycling process, from conversion to MOX reprocessing. Several scenarios for disposition of Weapons-Grade Pu were evaluated, depending on the type and number of reactors to be involved. As far as MOX fabrication is concerned, scientists studied the feasibility of constructing a first facility (called TOMOX at that time) to incorporate Russian WG-Pu into MOX fuel for VVER 1000 and BN 600 reactors. Following this work, a feasibility study on TOMOX was released in December 1995 by COGEMA Group, and, since the program was to last four years, the final report on the AIDA MOX work was made public in March 1997.

On the German side, two major documents have been released by GRS / SIEMENS / MINATOM. The first that was issued in 1994 described a feasibility study for a MOX Pilot Plant and the second that detailed the basic design of the MOX plant was released in 1996. As a result of four years of bilateral German/Russian and French/Russian cooperation the different studies came to similar results:

- The loading of MOX fuel, made from WG-Pu into Russian VVER 1000 and fast reactors, especially the Balakovo units and the BN 600, is feasible.
- To fabricate this fuel, concepts for pilot scale fabrication plants were prepared: TOMOX from the French/Russian cooperation and the MOX Pilot Plant from the German/Russian cooperation.

3.2. COGEMA/SIEMENS PROJECT

In 1996, Russia, Germany and France have decided to combine their efforts in a joint initiative for collaboration in the field of the peaceful use of WG-Pu in Russia. Consequently, SIEMENS/COGEMA announced a joint project for a MOX fuel fabrication plant in collaboration with MINATOM. This plant will convert 1300 kg WG-Pu into MOX and FBR fuel to be loaded in Russian reactors after 2001. This project is called **DEMOX.** Further steps might include a larger MOX capacity able to treat up to 5 tons of WG-Pu per year. Operation experience in the MOX field gained by COGEMA (Cadarache and MELOX) and SIEMENS (Hanau) are combined in the DEMOX project.

The project involves a common engineering team working in COGEMA's offices (since March 1997) and led by a SIEMENS project manager. An 'harmonization phase' has already confirmed the design options of the project.

The results are that a MOX plant based both on COGEMA's Advanced-MIMAS process and on the Hanau plant design and existing equipment, is feasible. A draft report of this effort has been completed.

Common work with MINATOM is ongoing in order to prepare for the start-up of a common French-German-Russian Basic Design in the fall of 1997. All basic documents for deciding the DEMOX investment (basic design, cost calculation, safety report, financing scheme) will be prepared during 1997/98. All of this work is being done in accordance with the usual safety constraints and in compliance with IAEA safeguards. The preparation phase and the basic design are supported by French and German government funds until mid 1998. Construction of the DEMOX plant would follow until 2001 in parallel with some actions planned in the fields of MOX irradiation and reactor adaptation.

In summary, by the year 2001, the whole industrial system needed for fabrication and loading of MOX fuel made from WG-Pu could be available in Russia.

4. US Situation

4.1. WG-PU DISPOSITION IN THE US

The January 14th 1997 Record of Decision released by the DOE has endorsed a dual-track approach for US military plutonium disposition. Further studies will be conducted concerning Pu immobilization and MOX options for about two years. On July 17th, 1997, the DOE issued its Program Acquisition Strategy (PAS) for obtaining MOX fuel fabrication and reactor irradiation services. This document describes the way DOE intends to work with US and foreign industrial organizations to implement a MOX option. It enumerates the criteria under which the candidates will be evaluated. The Program Acquisition Strategy is open for comment by utilities and industry until the middle of September 1997.

4.2. COGEMA ACTIVITIES

Since the beginning of the US Pu disposition discussions, COGEMA has been closely following US discussions and successive reports released by DOE. In January 1996, the company was one of the industrial organizations to respond to DOE Request for Expressions of Interest in a MOX program. COGEMA also supported the US utilities, Duke Power and Commonwealth Edison, in their action to promote the MOX route through P.E.A.C.E. Project and provided them with information that they needed concerning MOX fuel technology and experience. Many discussions are ongoing between COGEMA and DOE, as well as with US utilities and industrial organizations while waiting for the Request For Proposal on MOX fabrication and irradiation services expected in February 98. Some comments have been sent to DOE on the Program Acquisition Strategy released in July and COGEMA is ready to participate in a consortium made up in accordance with DOE expectations.

5. Conclusions

The military Pu disposition program can benefit from the civilian MOX industry's experience and know-how. MOX has become a well proven technology that is now use in several European countries and is soon going to be introduced in Japan. COGEMA Group has recognized experience in this field and could make its expertise available for WG-Pu recycling programs. Some co-operation programs have already started. As far as the industry is concerned, since no technical issue is anticipated, the implementation of such a program could start soon. However it is clear that such programs will need a strong political initiative to overcome institutional and financial issues.

PLUTONIUM INCINERATION IN LWRs

The First Step in the Double Strata Fuel Cycle

J. MAGILL
H. J. MATZKE,
J. VAN GEEL
European Commission, Joint Research Centre,
Institute for Transuranium Elements
Postfach 2340,
D-76125 Karlsruhe,
Federal Republic of Germany.

Abstract

Much attention has been paid recently to the present and future build-up of plutonium stocks and the various "Bury or Burn" strategies[1] for management of this material. The plutonium arises from the dismantling of nuclear weapons (W-Pu) and from reprocessing of spent fuel (C-Pu). In this paper, the incineration of C-Pu in various matrices--in urania as MOX, in thoria as TMOX and in inert matrices such as $MgAl_2O_4$ as IMOX -- is considered. This constitutes a first fuel cycle stratum based on LWRs and once through recycling with a view to decreasing the Pu stockpile in spent fuel. In the second stratum which is not the subject of this paper, minor actinides and long lived fission products are partitioned and then, transmuted in a dedicated actinide burner.

1. Introduction

The global total nuclear power generating capacity at the end of August 1996 was $361GW_e$[2]. Assuming this power generating capacity remains constant, one can estimate the annual production rate of plutonium in spent fuel (this plutonium is produced through neutron capture reactions beginning with ^{238}U in standard fuel). Under normal operation, the specific energy content of PWR fuel is 33 $GW_{th}d/MTHM$ (Gigawatt days per metric tonne of heavy metal). Hence 361 $GW_e y$ results from 1.14×10^4 tonnes of fuel (assuming all reactors to be PWRs with a discharge burnup of 33 $GW_{th}d/MTHM$ and a 35%

T. A. Parish et al. (eds.),
Safety Issues Associated with Plutonium Involvement in the Nuclear Fuel Cycle, 91–95.
© 1999 *Kluwer Academic Publishers. Printed in the Netherlands.*

conversion efficiency from thermal to electrical energy). From Table 1 it can be seen that approximately 0.9% of the spent fuel mass is in the form of plutonium isotopes. This implies a global production of approximately 100 tonnes of plutonium per year. In reality, only about half of the spent fuel originates from LWRs. The rest arises from other reactor systems such as CANDU. Since these systems generate less Pu, the above result is only approximate and represents an upper limit. If no recycling of the spent fuel occurs, by 2050, this will lead to a
stockpile of approximately 5000 tonnes of plutonium in spent fuel in addition to the Pu already existing from reactor operation before 1997.

To reduce this build-up of plutonium, the spent fuel can be reprocessed and the plutonium recycled back into the reactors (by the year 2000, however, the reprocessing capacity for spent fuel is expected to result in only about 20 tons Pu per year for use in MOX). This is the power reactor part of the so-called dual strata fuel cycle[3]. In the first stratum, the uranium and plutonium is multiply recycled. In the present paper, however, we consider the so-called ORTO (Once Recycled Then Out) process[4]. In the second stratum, the remaining minor actinides and long lived fission products are partitioned, and transmuted in a dedicated actinide burner. In this scenario, the reprocessing plant, partitioning and transmutation plants are co-located in a high level waste management park.

2. Calculations

In the following calculations, the recycling and incineration of plutonium is considered for various host matrices. The calculations were made using the ORIGEN2 computer code using appropriate cross section libraries[5]. In Table 1, the first row shows the amounts of plutonium generated (through capture reactions in ^{238}U) in PWRs at normal (33 GWd/MTHM) and extended (50 GWd/MTHM) burnups. Approximately 1% of the initial loading is in the form of plutonium isotopes. This is denoted as reactor-grade plutonium, ie., C-Pu, to distinguish it from weapons-grade, ie., W-Pu, which has a different isotopic composition. The C-Pu which is generated may be separated from the spent fuel by reprocessing and used to make new fuel. Hence, instead of using UO_2 fuel enriched with ^{235}U, the UO_2 is "enriched" with C-Pu to form mixed oxide (MOX) fuel. The results of burning MOX with a C-Pu enrichment of 5% in PWRs is also shown in Table 1. An initial mass loading of 50 kg of C-Pu per tonne of fuel (5%), is reduced to 36 (3.6%) and 32 (3.2%) kg per tonne at burnups of 33 GWd/MTHM and 50 GWd/MTHM, respectively. Also shown in Table 1 are the results for burning the C-Pu in an inert matrix (IM) at 5% enrichment (density of matrix assumed to be the same as that of Pu) and in a thoria matrix at 7.5% enrichment. One is, however, not free to choose the degree of enrichment of the C-Pu in the host matrix. Neutron economy should be as close as possible to that of just above 1.3, using standard fuel (3.2% ^{235}U enriched) in the PWR. Starting from an initial value for the infinite neutron multiplication factor, k_{inf} decreases with burnup to a value just less than 1.0 after 879 days (33 GWd/MTHM) irradiation.

TABLE 1. Total plutonium content (w/o) in fresh and spent PWR fuel at normal and extended burnups for various fuel configurations

fuel configuration	Fresh	33GWd/MTHM	50GWd/MTHM
UO_2 (3.2% enriched)*	0	0.89%	1.04%
MOX (5% C-Pu)*	5.0%	3.6%	3.2%
TMOX (10% C-Pu/ThO_2)*	7.5%	4.5%	3.4%
IMOX (5%C-Pu/IM)**	5.0%	1.7%	1.1%

*Power rating is 37.5 MW/tonne, **Flux is 2×10^{14} n/cm^{-2}s^{-1}

Our choices for the alternative matrix and degree of C-Pu enrichment need to lead to a variation in k_{inf} with burnup as close as possible to that of the reference case (i.e. 3.2% ^{235}U enriched fuel). This can be achieved by using MOX, IMOX (enriched with 5% C-Pu), and TMOX (enriched with 7.5% C-Pu). Similar calculations show that this is not the case for TMOX with 5% C-Pu. In addition, MOX and IMOX cannot be used with 7.5% Pu content without the introduction of a neutron poison such as Gd_2O_3.

3. Conclusions

It follows from Table 1 that for a burnup of 33 GWd/MTHM, the MOX fuel (5% enrichment) can burn 14 kg C-Pu per tonne of fuel (155kg Pu/GW$_{th}$y) and thoria fuel can burn 30 kg C-Pu (332kg Pu/GW$_{th}$y). At an extended burnup of 50 GWd/MTHM, the MOX fuel burns 18 kg C-Pu (131kg Pu/GW$_{th}$y) and the thoria fuel burns 41kg (299kg Pu/GW$_{th}$y). These values are consistent with the fact that the fissioning of 1 kg of Pu results in approximately 1 GW$_{th}$d of energy (more exactly 0.88 GW$_{th}$d) and this implies an *upper limit of 414 kg Pu fissioned per GW$_{th}$y*. Hence, if the present global nuclear power generating capacity of 361 GW$_e$ were fueled purely by fissioning Pu in LWRs, 426 tons Pu would be fissioned per year.

To counteract the production of Pu in standard LWRs, we need to dedicate a fraction of the existing reactors (i.e. in the first stratum of the fuel cycle) to use (ideally) full MOX, IMOX, or TMOX cores for Pu incineration or to have dual assemblies. As an example, consider a burnup of 33 GW$_{th}$d/MTHM at which the spent fuel contains approximately 1% C-Pu (see Table 1). Depending on the degree of enrichment, ξ, of the C-Pu in the MOX, IMOX, or TMOX, the fraction of power produced by standard fuel is approximately $\xi/(1+\xi)$ for the ORTO (once recycled then out) process. Similarly the fraction of power produced by MOX, IMOX, TMOX is $f_{XMOX} \cong 1/(1+\xi)$. The results are given in Table 2.

94

TABLE 2. Results of using a dual assembly or dual reactor system on the net Pu production rate

Dual assembly or reactor system	Pu from UO_2(tons/y)*	f_{XMOX}	Net Pu (ton/y)
UO_2	+101.5	0.0	101.5
UO_2 /MOX (5% C-Pu)	+86	0.15	62
UO_2 /TMOX (7.5%C-Pu/ ThO_2)	+90	0.11	54
UO_2 /IMOX (5%C-Pu/IM)	+86	0.15	29

*at 33GWd/MTHM, Pu balance calculated from Table 1 assuming total power of $361GW_e$.

It can be concluded that although significant amounts of energy can be obtained by burning Pu in LWRs, the annual net build-up can at most be decreased by a factor 3.5 (using IMOX) assuming a fuel burnup of 33 GW_{th}d/MTHM. The assumption made is that all spent UO_2 fuel is reprocessed once to separate the plutonium. This plutonium is then recycled as MOX (with urania, thoria or inert matrix host). Further reprocessing is significantly more difficult due to the higher heat, gamma, and neutron activities and will result in only a small additional reduction in the amount of plutonium[6]. A further reduction in the actinide inventory requires dedicated actinide burners.

References

1. Van Geel, J., Matzke, Hj., Magill, J. (1997) Bury or Burn? Plutonium - The Next Nuclear Challenge Nuclear Energy **36** 305-312.
2. World Nuclear Industry Handbook (1996), Nuclear Engineering International, p.9
3. Mukaiyama, T., Ogawa, T., Mizumoto, M., Takizuka, T., Hino, R., Oyama, Y. (1997) Omega Progam & Neutron Science Project for Development of Accelerator Hybrid System at JAERI, Proceedings of the I.A.E.A. Technical Committee Meeting on the Feasibility and Motivation for Hybrid Concepts for Nuclear Energy Generation and Transmutation, Madrid, Spain 17-19 Sept. 1997 to be published.
4. Magill, J., Peerani, P., van Geel, J., Rief, H., Wider, H. (1997) Nuclear Waste Incineration - The Case for ADS, Proceedings of the I.A.E.A. Technical Committee Meeting on the Feasibility and Motivation for Hybrid Concepts for Nuclear Energy Generation and Transmutation, Madrid, Spain 17-19 Sept. 1997 to be published.
5. Croff, A. G. (1983) ORIGEN2: A Versatile Computer Code for Calculating the Nuclide Composition and Characteristics of Nuclear Materials, Nucl. Tech. **62** 335-353. For the thorium calculations, the library pwrputh.lib for ThO_2 enriched with C-Pu was used. For the inert matrix and MOX calculations, the library prwus.lib for a standard 3.2% enriched uranium PWR was used.

6. Baetsle, L. H., De Raedt, Ch. (1997) Limitations of actinide recycle and fuel cycle consequences: a global analysis. Part 1: Global fuel cycle analysis, Part 2: Recycle of actinides in thermal reactors: impact of high burn up LWR-UO$_2$ fuel irradiation and multiple recycle of LWR-MOX fuel on the radiotoxic inventory. Nuclear Engineering and Design **168** 203-210.

STATE OF THE ART AND OUTLOOK FOR THE NUCLEAR FUEL CYCLE IN THE RUSSIAN FEDERATION

M.I. SOLONIN
A.S. POLYAKOV
B.S. ZAKHARKIN
*SSC RF A.A. Bochvar's All-Russia Scientific Research Institute
of Inorganic Materials, Moscow, Russia*

1. Introduction

Nuclear reactors which act to generate power from fission are only one element in the sophisticated complex of operations that compose the nuclear "fuel cycle". The efficient production of nuclear power from uranium fuel involves a series of steps including mining and milling, conversion and enrichment, fabrication and radiochemical reprocessing. The majority of these processes use advanced technologies, and their exploitation is the prerogative of individual countries in the world. Currently, there are 447 commercial nuclear reactors operating worldwide in 31 different countries. However, the full range of commercial fuel cycle processes have been fully developed primarily in only three countries, the United Kingdom, France and the Russian Federation.

Russia has been one of the world's leaders in pioneering the development of the sophisticated technologies needed in the nuclear field. While solving urgent defence related problems, it has created in a short time a powerful complex of nuclear fuel cycle facilities that can be fully employed for peaceful nuclear power applications.

The nuclear fuel cycle is divided into two main parts, namely, the "front end" and the "back end." The "front end" of the fuel cycle consists of three parts -- mining and milling, conversion and enrichment, and fuel assembly fabrication. After these steps are completed, the fuel assemblies are used to produce power in nuclear powerplants. Following discharge as spent fuel from nuclear powerplants, the assemblies enter the "back end" of the nuclear fuel cycle which also consists of three parts -- storage, reprocessing and waste disposal. Russia has significant experince with both the "front end" and "back end" processes of the nuclear fuel cycle.

T. A. Parish et al. (eds.),
Safety Issues Associated with Plutonium Involvement in the Nuclear Fuel Cycle, 97–104.
© 1999 *Kluwer Academic Publishers. Printed in the Netherlands.*

2. Fuel Cycle Steps

The first step in the nuclear fuel cycle is mining and milling. The endproduct of milling is purified uranium oxide, the basic stock material for nuclear powerplant fuel. Uranium resources in known deposits in the former USSR are 2 million tons which corresponds to the amount of fuel needed to supply the needs of about 300 GW_e of nuclear power plants over their lifetimes. For the most part, uranium deposits were developed in Middle Asia, but Russia accounts for some 30% of the mining capacity. Today, the uranium stock available is sufficient not only to supply all of the fuelling needs of domestic and foreign nuclear powerplants of Russian design, but also to actively participate in the world export market.

The second step in the nuclear fuel cycle is conversion of the uranium oxide to uranium hexafluoride followed by isotopic enrichment. In Russia, there are production plants engaged in conversion and enrichment which have capacities to supply the refuelling needs of 100 GW_e of nuclear powerplants. The enrichment plants which are used to increase the U-235 content (isotopic enrichment) of the natural uranium hexafluoride product make use of high efficiency centrifuge technology. The world leading cost effectiveness and reliability of its enrichment facilities have allowed Russia to successfully enter the world market for uranium separation services (France is among the customers).

Russia also has significant facilities and experience in nuclear reactor fuel fabrication, the third step in the nuclear fuel cycle. Russian plants produce fuel pellets, claddings, spacers and assemblies of different kinds. The quality of nuclear fuel assemblies fabricated in Russia has been verified to exceed exacting internationally accepted standards. By cooperating within the framework of the CIS, fuel fabrication facilities are available with a capacity to meet the full demand of nuclear powerplants with a generation capacity totalling 120 GW_e.

In terms of the back end of the nuclear fuel cycle, Russia also has significant facilities and experience. The radiochemical complex (RT-1 plant) has for some 20 years carried out reliable and safe reprocessing of spent fuel rods from large and small power reactors (VVER-440, BN-600, fuel rods from ice-breaker and submarine reactors, and research reactors). Some nuclear fuel from outside countries has also been reprocessed (see Figs. 1 and 2). Up to the present time, the RT-1 plant has reprocessed some 3500 t of spent fuel of Russian origin.

In principle, spent fuel reprocessing is an important link in the nuclear fuel cycle. The alternatives of reprocessing or not reprocessing determine whether the fuel cycle is open or closed. From the very start of nuclear power generation in Russia, the intention has been to completely close the fuel cycle (Fig. 3) in order to fully exploit the energy potentially available from uranium. Closing the commercial nuclear fuel cycle with reprocessing depends greatly on two factors: cost effectiveness and safety.

The potential cost-effectiveness of reprocessing is based on recovering and recycling both the unburnt uranium and plutonium extracted from spent fuel. Both elements are obtained from chemical separation processes applied to irradiated fuel. The potential for improved safety as a result of reprocessing is due to 1) a reduced demand

Fig. 1. Location of the RT-1 Plant

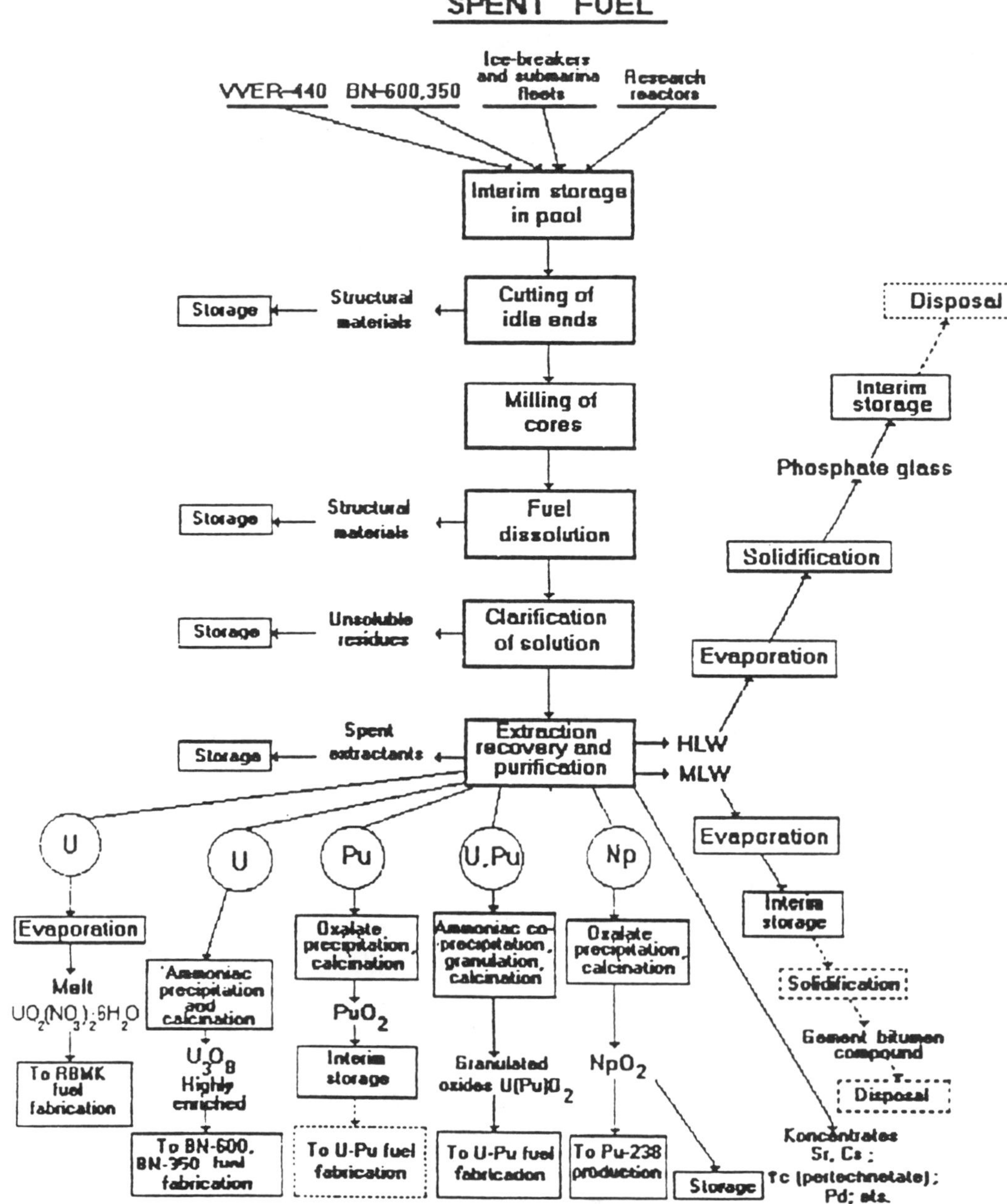

Fig. 2. Functional Processes of the RT Plant (dashed boxes denote process steps that are underconstruction or justification (waste disposal))

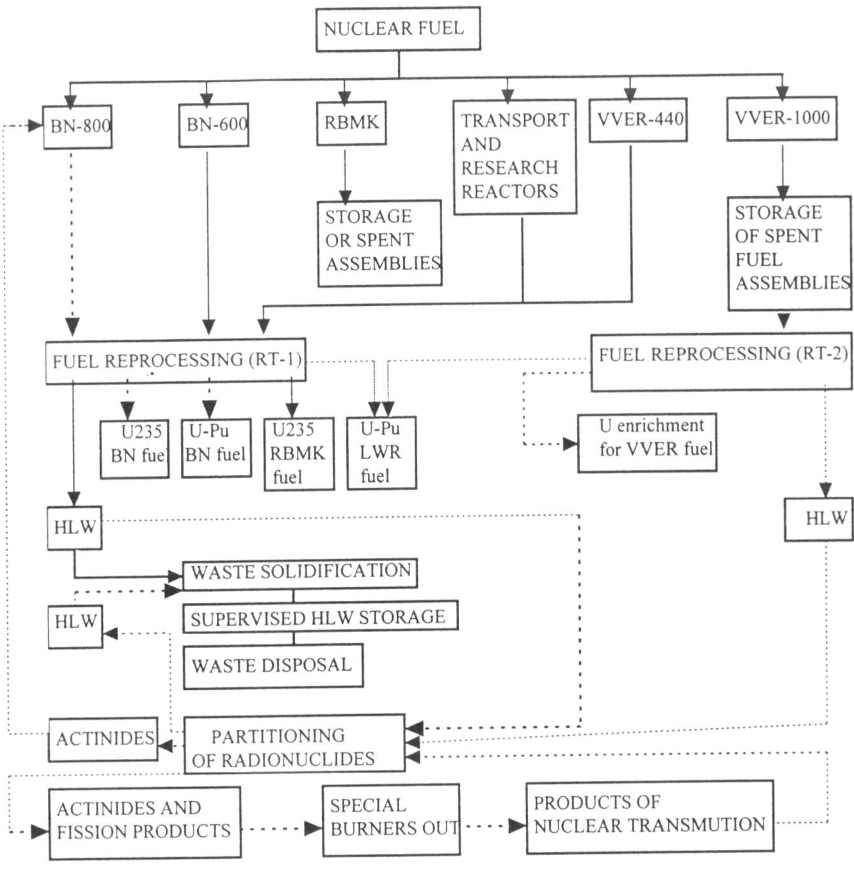

Fig. 3. Schematic Diagram of the Nuclear Fuel Cycle in Russia

for natural uranium, and therefore, reduced ecological damage from mining and milling operations and 2) burning up plutonium and other potentially dangerous actinides in spent fuel. Consumption of plutonium by recycle to reactors produces energy from this resource and reduces the amount of long-lived isotopes that are considered to be radioactive waste, and that therefore, have to be incorporated into stable matrix material forms for subsequent geological disposal.

3. Closing the Fuel Cycle

Due to the successful operation of the RT-1 plant, a decision was made to go ahead with recycling of the recovered uranium, in other words, the closing of the fuel cycle in this respect. A commercial process has also been mastered to solidify the high level radioactive wastes (Fig. 4) produced from reprocessing. Since 1987, some 290 million Ci of high level radioactive waste has been incorporated into phosphate "glass". In addition, 30 t of plutonium has been extracted from civilian reactor fuel and is being stored for future non-military use.

In support of future plutonium recycle to nuclear powerplants, research and development on a large scale have been implemented to evaluate the technologies needed for practical U-Pu fuel fabrication. Some of these technologies have been demonstrated on a semi-commercial scale at the "Mayak" and RIAR facilities. There U-Pu fuel has been fabricated in an amount of more than 1 t. Most of this fuel was tested to high burnups in the fast reactors BN-350, BOR-60 and BN-600. In the 1980s, construction of a special complex in the Urals region was started to fabricate U-Pu fuel for BN-800 reactors.

The economic difficulties of the last decade have drastically slowed down the implementation of the Russian reactor programs, and have completely terminated the construction of the needed fuel cycle related facilities. However, it goes without saying that, as electricity demand grows, the need to implement a fully functioning nuclear fuel cycle will eventually dictate the renewal of the interrupted work. In the near term, priority is likely to be given to the reprocessing industry for the VVER-1000 spent fuel that is quickly filling the storage areas at nuclear powerplants. An alternative to the RT-2 plant, the construction of which was stopped at its initial stages, is foreseen to be a much lower expense option that consists of incorporating new head-end units that will allow the large scale reprocessing of VVER-1000 fuel within the structure of the RT-1 plant. The advisability of this option is also enhanced by the fact that the RT-1 plant's capacity, in terms of the VVER-440 fuel recently processed, is utilized to less than 20%.

The renewal of construction work and eventual completion of the complex to fabricate U-Pu fuel must be coordinated in time with the deployment of BN-800 units. The construction of the U-Pu fabrication complex will require a substantial lead time and its operation will need to be coordinated so that fuel assemblies are finished about one year prior to their loading into reactors. The technological "fulfillment" of the fuel complex will require an improved design taking into account the results of research and development and semicommercial tests carried out recently.

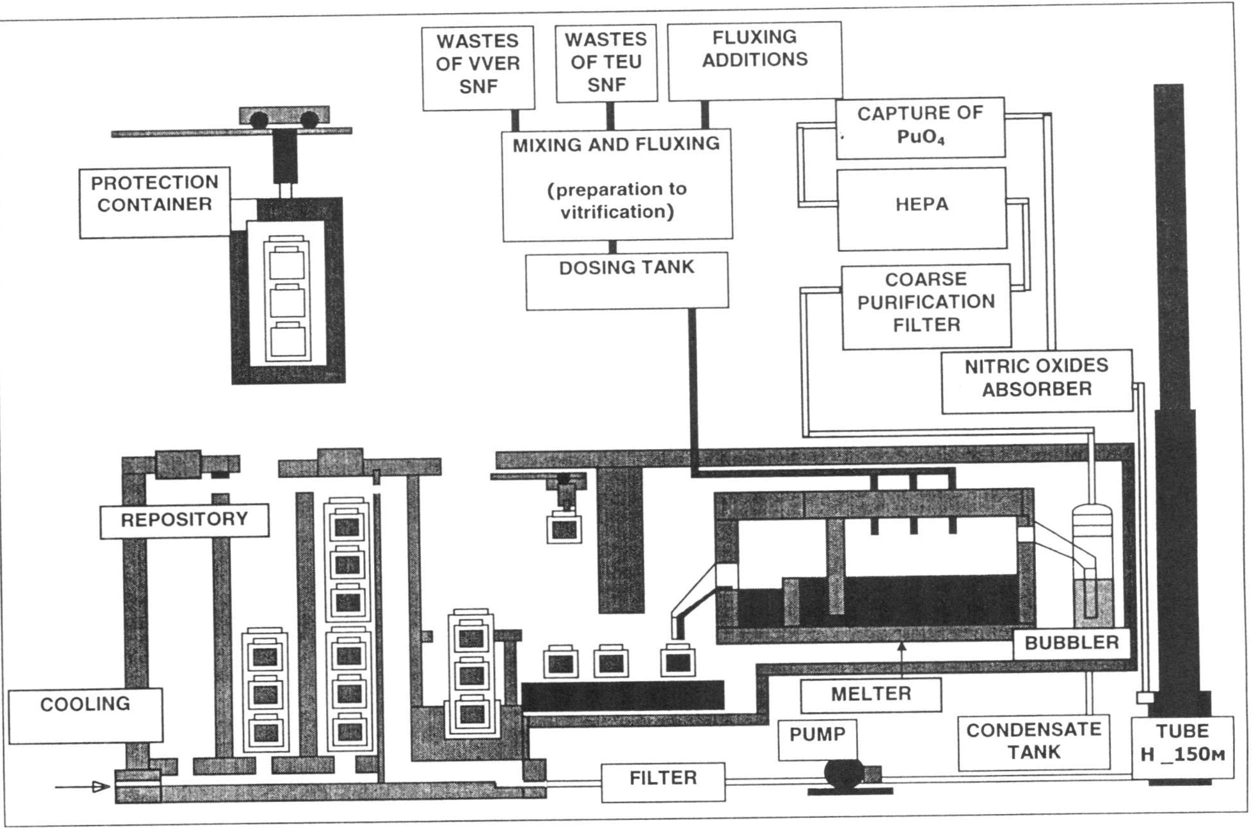

Fig. 4. Flow-sheet of HLW Vitrification Using EP-500 Commercial Melter

4. Conclusions

Among the urgent research and development needs related to the fuel cycle processes are projects that deal with the utilization of radiologically dangerous long-lived nuclides, primarily, the minor actinides, Np, Am, and Cm. Process developments aimed at partitioning these elements from spent fuel in combination with the nuclear transmutation of the target long-lived species in either conventional nuclear reactors, or specially designed "burner" reactors, may make it possible to maximally decontaminate the final waste from the nuclear fuel cycle before commiting it to disposal in geological strata. Also promising are investigations into the exploitation of the U-Th fuel cycle. This fuel cycle has huge known fuel reserves and the initial activity level of spent U-Th fuel is 10^2 to 10^3 times lower than that from the U or U-Pu fuel cycles. The U-Th fuel cycle will not compete soon on a practical scale with the U or U-Pu fuel cycles. However, owing to its large fuel reserves, the generation of scientific knowledge in this area is advisable at the present time.

ON-SITE SPENT FUEL MANAGEMENT BASED ON ELECTROREFINING

V. A. KHOTYLEV
D. R. KINGDON
A. A. HARMS
J. E. HOOGENBOOM
*Department of Engineering Physics, McMaster University
Hamilton, Ontario L8S 4L7 Canada
*Interfaculty Reactor Institute
Delft University of Technology
Mekelweg 15, 2629JB, Delft
The Netherlands*

Abstract

The use of weapons grade fissile materials for fuel make up is considered in the context of a new strategy for closure of the nuclear fuel cycle. The reduced volume, activity, and lifetime of radioactive waste in a strategy that incorporates on-site electrorefining of the spent fuel is calculated and compared to the more conventional once-through fuel cycle scheme. Significant reductions in the waste activity and the duration of its existence by recycling all but neutron-absorbing fission products back into the reactor core allows for the possibility of on-site storage as an alternative to permanent disposal without isolating plutonium nor introducing any attendant increased proliferation risk.

1. Introduction

The investigations of spent nuclear fuel management have received a great deal of attention from both the nuclear industry and the public. Several concepts to close the nuclear fuel cycle through burial of waste in deep, stable geological formations have been examined, but no facility has yet been provided which is acceptable to all those involved. Herein, another approach is examined in which electrorefining is used to remove neutron-absorbing fission products from the spent fuel, all other species being recycled back into the reactor core for subsequent burnup or transmutation to more stable isotopes. Because this type of reprocessing requires that only the extracted fission

105

T. A. Parish et al. (eds.),
Safety Issues Associated with Plutonium Involvement in the Nuclear Fuel Cycle, 105–109.
© 1999 *Kluwer Academic Publishers. Printed in the Netherlands.*

products be disposed of, the volume of waste, its radioactivity and its lifetime are significantly reduced, potentially allowing for on-site storage as an adequate means for closing the fuel cycle. Also significant is the fact that the process does not involve the isolation of plutonium, and thus avoids the associated risk of nuclear proliferation.

2. On-site Strategy

Central to the spent fuel management strategy considered here is the electrorefining technique which removes some or all of the neutron-absorbing fission products, but retains the rest of the fuel -- notably the remaining fissile and transuranic materials -- for use in the manufacture of new fuel pellets. Since the main long-term (>500-1000 years) contributors to the radiological hazard of spent nuclear fuel are the minor actinides and other transuranic elements, due to their long half-lives and large decay heats, their recycling back into the core, and thus removal from the waste stream means that the isolation barriers associated with a disposal concept are no longer required to achieve insolubility and immobility for 10000-100000 years, or longer. Instead, only the relatively short-lived (<500 years) fission products require disposal, reducing the time scale for the integrity of the isolation barriers to remain intact, and thus easing the requirements on any type of disposal concept.

Electrorefining processes have been developed over a number of years [1,2]. They use high temperature (500°C) molten-salt and molten-metal solvents to electrochemically remove over 99.9% of the U, Pu, and other transuranic elements from the spent fuel. All these isotopes are removed collectively as one medium, and thus there is no isolation of Pu with its attendant proliferation risks. The operation takes place in a heavily shielded hot-cell facility, and begins by chopping up the spent fuel assemblies into small pieces (~6-7 mm) which are placed in a steel basket that acts as the anode in the electrochemical cell. A voltage of ~1 V is applied between this anode and one of several cathodes; a steel cathode collects essentially pure uranium, while uranium, neptunium, plutonium, americium, curium and some rare-earth fission products are collected at a liquid cadmium cathode. The majority of the fission products -- alkali metals, alkali-earth metals and rare-earths -- are left behind in the electrolyte salt, while structural materials and a few other fission products (i.e. noble metals) remain in the molten cadmium or the anode basket [3,4]. Fission product gases like xenon and krypton, and other gases such as tritium are recovered from the argon over-gas following the electrolysis and are stored for decay.

The separated media from the spent fuel are removed in batches from the electrolytic bath for purification and to collect and prepare the fission products for storage. The cathode deposits are melted in a high temperature furnace to evaporate the cathode materials and any other impurities, leaving behind metal ingots ready to be made into new fuel. All components of this fuel recycling, waste disposition, and new fuel assembly manufacture system have been built and tested -- some at full scale -- for a metal-fueled reactor [2,3].

The same process is possible for non-metal fueled reactors by adding some preliminary steps to the electrorefining operation. In the case of light water reactor fuel, the zircalloy cladding is first removed, followed by reduction of the oxide fuel to metallic form by a lithium reactant. The remainder of the fission product extraction process is as outlined above, and finally the lithium reactant is recovered by electrolysis prior to the manufacture of new fuel elements [5].

3. Assessment

To assess the merits of such a spent fuel management strategy, a comparison of the radioactive waste activity as a function of time derived from such an operation made to that of a simpler once-through fuel cycle (without reprocessing). Essentially, the activity for the materials extracted during the electrorefining operation has been calculated for a PWR, subject to typical operation and fuel burnup scenarios.

Consider a unit volume from within a reactor core that is initially loaded with fresh fuel. Following operation for ~3 years, the majority of the fission products are removed and makeup fissile fuel is added. Any remaining fission products and all of the actinides created during the burnup period remain in the volume considered. The burnup cycle is repeated, and again removal of the most neutronically poisonous isotopes and top-off with fresh fuel occurs. The procedure is repeated, and the accumulation of activity from the waste removed between each burnup cycle is calculated. Comparison of this activity accumulation with that of removing the entire volume of fuel following each cycle provides a measure for the reduced amount of radioactive waste generated by employing this spent fuel strategy.

Burnup assessments were made with a new code package that utilizes functional modules of the SCALE 4.3 calculational system [6], which is distributed by RSICC and recognized by the NRC as valid for licensing spent fuel packages. Initially, neutronic calculations were used to generate cross-section libraries, first for individual fuel elements and subsequently for entire fuel assemblies, considering about 200 distinct nuclides. These libraries are then used along with the initial material concentrations to assess the burnup of 750 light isotopes, more than 100 actinides, and more than 1000 fission products over the duration of the burn period. During each burnup period an updated neutron spectrum is calculated from intermediate isotopic concentrations at several stages so that each cycle is divided into at least 3 subsections, each with its own spectrum and libraries to provide an accurate assessment of the material composition changes with burnup.

As a typical unit volume, one fuel assembly and its surrounding structure, coolant, moderator and control media were chosen for use in these calculations. The power density was held constant at values representative of the average power density in a PWR, and criticality was maintained by adjusting a poison concentration in the volume itself.

Following each burnup period, all gases -- fission product or otherwise -- are removed as they are released in the electrorefining procedure. The fission products' absorption is characterized by flux-averaged macroscopic cross-sections, the least radiologically hazardous elements are removed until sufficient volume and negative reactivity has been extracted to allow fresh fuel to make up for the reactivity deficit. Of

108

course, the total volume of the assembly considered must remain the same at the beginning of each burn cycle.

Calculational modeling has shown that the accumulation of radioactivity per unit power produced, and the lifetime thereof, compared to a conventional system is

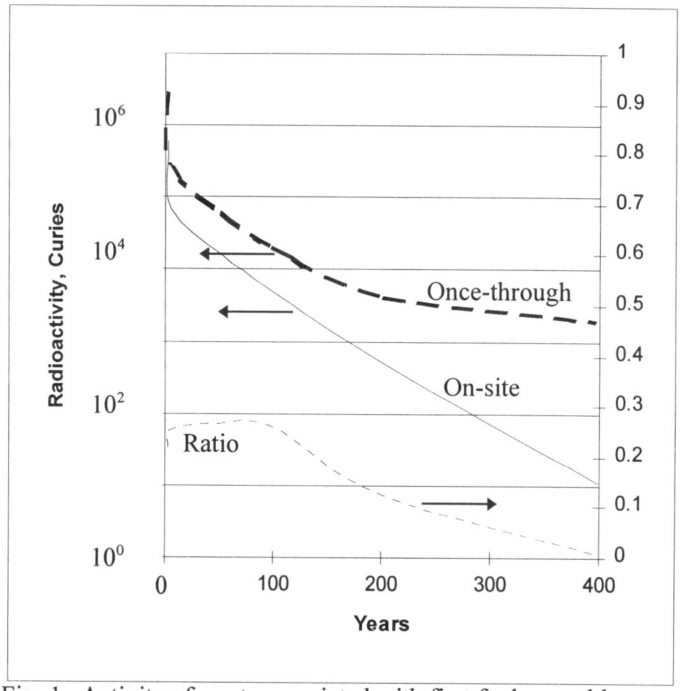

Fig. 1. Activity of waste associated with first fuel assembly.

significantly reduced. The volume of the wastes does not become prohibitively large over time, even if the operation were to last over several decades or even centuries.

For a continuously operating system with storage of the wastes on site, the activity ratio at the time of each batch's removal from the reactor was found to be the determining factor in the extent of waste activity reduction, see Figure 1. This is because the relatively short-lived fission product species are the only ones of relevance in the waste stream, and thus each addition to the waste collection becomes the major component thereof. The long-term effects are instead, primarily determined by the activity ratio after an extended length of time. The effect of the nuclear composition of the newly added fuel, i.e. low or high enrichment uranium, mixed uranium and plutonium, or plutonium alone on the degree of waste activity reduction was also investigated.

4. Conclusions

On-site spent fuel management, in which electrorefining is used to remove fission products from spent fuel with all other species recycled back into the reactor for subsequent burnup or transmutation to more stable isotopes, appears to hold significant promise. Calculations have demonstrated that the volume of waste, its radioactivity and its lifetime are significantly reduced, thereby allowing for residual spent-fuel on-site storage as a promising prospect.

References

1. Laidler, J. J., et. al. (1997) Development of Pyroprocessing Technology. *Progress in Nuclear Energy* **31**, 131-140.
2. Koyama, T., et. al. (1997) An Experimental Study of Molten Salt Electrorefining of Uranium Using Sold Iron Cathode and Liquid Cathode for Development of Pyrometallurgical Reprocessing. *Journal of Nuclear Science and Technology* **34**, 384-393.
3. McFarlane, H. F., and M. J. Lineberry. (1997) The IFR Fuel Cycle Demonstration. *Progress in Nuclear Energy* **31**, 155-173.
4. Chow, L. S., et. al. (1993) Continuous Extraction of Molten Chloride Salts with Liquid Cadmium Alloys. *Proceedings of the International Conference on Future Nuclear Systems: Emerging Fuel Cycles and Waste Disposal Options*, 12-17 September 1993, Seattle, Washington. Published by the American Nuclear Society.
5. McPheeters, C.C., et. al. (1997) Application of the Pyrochemical Process to Recycle of Actinides from LWR Spent Fuel. *Progress in Nuclear Energy* **31**, 175-186.
6. *Modular Code System for Performing Standardized Computer Analysis for Licensing Evaluation for Workstations and Personal Computers SCALE 4.3*, (1995) RSICC Computer Code Collection, CCC-454.

PROSPECTS FOR IMPROVEMENT OF VVER FUEL MANAGEMENT AND MOX UTILIZATION

A. N. NOVIKOV
V. I. PAVLOV
A M. PAVLOVICHEV
V. N. PROSELKOV
V. V. SAPRYKIN
I. K. SHISHKOV
Russian Research Centre
"Kurchatov Institute"
Kurchatov Square, 1
123182 Moscow
Russian Federation

1. Introduction

Extensive fuel operational experience has been gained at 20 VVER-1000 reactors (cumulative loading of ~7000 fuel assemblies) and 28 VVER-440 reactors (cumulative loading of ~18000 fuel assemblies) with maximum burnups to 49-50 GWD/MTU. Further increases in burnup are anticipated with the implementation of four year fuel residence times at VVER-440 units and three year fuel residence times at VVER-1000 units using low leakage loading patterns. Higher burnups will also be achieved with the introduction of both zirconium spacer grids and zirconium control rod guide tubes in the fuel assemblies.

Various ways of using weapons-grade plutonium in different types of power reactors have been investigated in Russia. However, an analysis that takes into account the current structure of nuclear power deployment in Russia reveals that VVER-1000 reactors, as well as VVER-440 reactors, appear to be the most promising for this purpose. Up to the present, Russian VVER deployment has totaled 28 VVER-440 power units and 20 VVER-1000 power units. Improvements in the fuel performance of these reactors involves enhancement of both safety and efficiency of fuel utilization. Examination of

111

T. A. Parish et al. (eds.),
Safety Issues Associated with Plutonium Involvement in the Nuclear Fuel Cycle, 111–120.
© 1999 *Kluwer Academic Publishers. Printed in the Netherlands.*

112

the historic pattern for discharging spent fuel (Figs. 1 and 2) shows that the average discharge burnup rose to a maximum then has decreased with time. At present, there is intense work in Russia to modernize the fuel management strategies applied to both VVER-440 and VVER-1000 reactors. These activities are being conducted within the framework of OAO TVEL with participation by some additional research and design institutes [1-3]. For VVER-440 reactors, the modernization efforts include:

- Using fuel enrichment profiling within the fuel assembly cross section;
- Designing cores with minimal neutron leakage;
- Introducing assemblies with zirconium spacer grids;
- Achieving four and five year fuel residence times;
- Possibly using uranium-gadolinium fuel (UGF); and
- Evaluating ways to improve the value of the power coefficient of reactivity at the end of core life.

For VVER-1000 reactors, the modernization efforts include:

- Using new fuel assemblies with both zirconium spacer grids and zirconium control rod guide tubes (ZFA);
- Using uranium-gadolinium fuel (UGF);
- Increasing the amount of fuel loaded in the reactor by reducing the size of the axial hole inside the fuel pellets;
- Using core loading patterns that achieve low neutron leakage;
- Changing to three-year, and subsequently, four-year fuel residence times; and
- Improving the power coefficient of reactivity at the end of core life.

2. VVER 440 Fuel Cycles

With the transition to modernized fuel cycles, which is now under way, the efficiency of fuel utilization is expected to be significantly increased. For VVER-440 type units, for example, this can be seen in Table 1, where the main characteristics of fuel cycles based on three and five year residence times are given for cores containing 349 fuel assemblies. The values shown for the fuel assembly enrichments and burnable absorber loadings are based on assuming that in all of the fuel assembly types, zirconium spacer grids are used, and the thicknesses of the shroud tube walls are 1.5 mm and 2.0 mm, respectively, in the fuel assemblies that do not contain and do contain removable burnable absorber rods.

Fuel cycle variant #1 for VVER 440 units (see Table 1) requires annual loading of 78 fuel assemblies with a U-235 enrichment of 3.6% that contain 12 burnable (and removable) neutron absorber rods and 66 fuel assemblies with a U-235 enrichment of 4.4%. Profiling of the enrichment across either type of fuel assembly's cross section is not used. The fuel assembly reloading pattern that is employed is a partial in-in-in-in-out strategy (this fuel cycle variant can be considered as an analog of that now used at the Kola Nuclear Power Plant Unit 3).

Fuel cycle variant #2 (see Table 1) requires annual reloading of 78 fuel assemblies with a U-235 enrichment of 3.6% containing 12 burnable (and removable) neutron absorber rods and 66 fuel assemblies with profiling of the fuel enrichment across

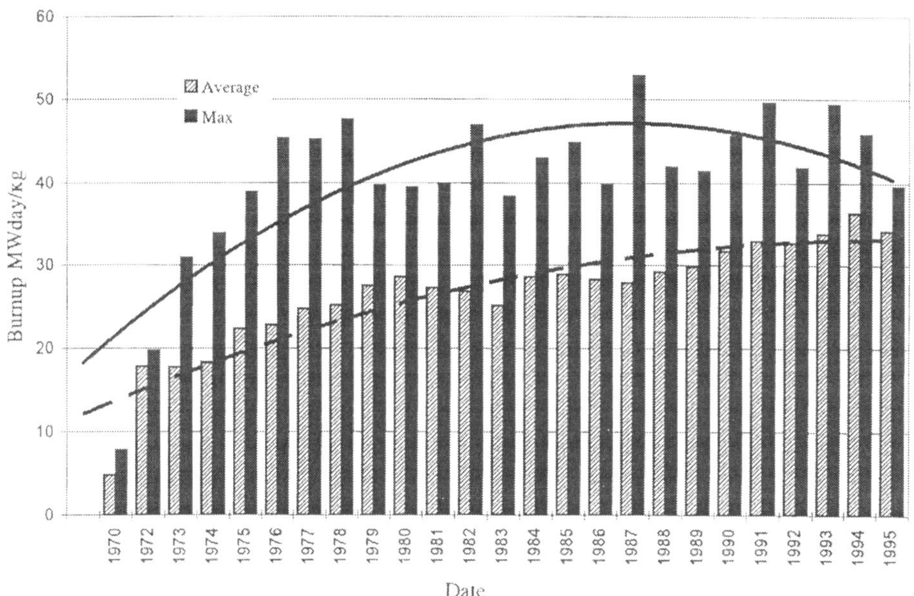

Fig. 1. Dependence of fuel burnup on the date of unloading for VVER-440 FAs

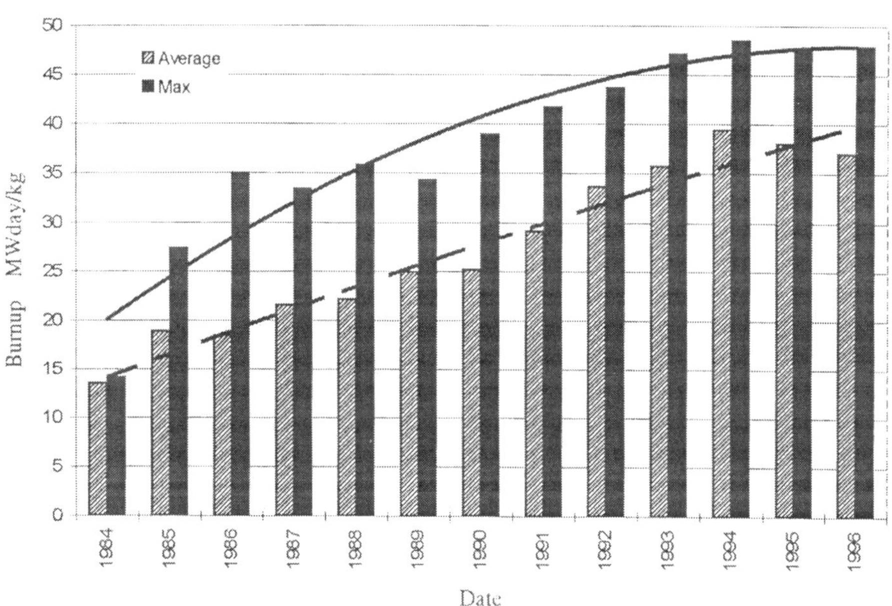

Fig. 2. Dependence of fuel burnup on the date of unloading for VVER-1000 FAs

the assembly cross section. The average fuel enrichment in the 66 assemblies with enrichment profiling is 4.2%. The fuel assembly reloading pattern for VVER-440 fuel cycle variant #2 is a complete in-in-in-in-out one.

Fuel cycle variant #3 (also see Table 1) requires annual reloading of 78 fuel assemblies with a U-235 enrichment of 3.6% containing 12 burnable (and removable) neutron absorber rods and 66 fuel assemblies that contain gadolinium burnable poison integrated into the fuel. Fuel enrichment profiling is used in these 66 assemblies and their average fuel enrichment is 4.41%. The fuel assembly reloading pattern for VVER-440 fuel cycle variant #3 is also a complete in-in-in-in-out scheme.

3. VVER-1000 Fuel Cycles

The characteristics of the modernized fuel cycles for VVER-1000 type reactors are listed in Table 2. In both fuel cycles with three-year and four-year fuel residence times, gadolinium burnable poison is used. An initial Gd loading of 5 weight percent is used in about 700 fuel rods for the three year residence time fuel cycle and in about 360 fuel rods for the four year residence time fuel cycle. Comparison of the characteristics of the modernized fuel cycles with those of the standard fuel cycles for the VVER-440 and VVER-1000 power plants shows considerable improvement in fissile fuel usage efficiency. In particular, conversion to the modernized fuel cycles permits:

- The consumption of natural uranium to be reduced by 12-19%;
- The total number of fuel assemblies used to be decreased by 25-30%, thus, reducing fuel fabrication costs, as well as, expenditures for transportation and storage of spent fuel; and
- The operational safety margins are increased, and the recriticality temperature is decreased.

In regard to fuel utilization, these changes result in performance for VVER reactors that is competitive with that of the modern designs of western PWRs.

As of the present, preliminary tests and demonstrations have been carried out to prepare for large-scale implementation of modernized VVER fuel cycles. In particular:

- A five-year fuel residence time reload scheme has been implemented at the Kola Nuclear Power Plant Unit 3;
- Four and five year fuel residence time reload schemes have been introduced at the Rovno Nuclear Power Plant Units 1 and 2;
- Fuel assemblies with profiled enrichment distributions are in use at the NV Nuclear Power Plant Unit 4;
- Experimental operation of UGF is being carried out at the Balakovo Nuclear Power Plant;
- Experimental operation of ZFA is being performed at the Balakovo, Rovno, and Zaporozhie Nuclear Power Plants; and
- Core designs with low neutron leakage are being used in many VVER units.

In addition to the above-mentioned methods for VVER fuel cycle modernization, the possibility of burning both weapons grade and reactor grade plutonium in VVER units has been considered. An analysis of nuclear power deployment and future power supply in Russia shows that the VVER-1000 reactors, as well as the VVER-640 reactor now under development, appear to be the most promising reactors for this mission in the short term. Studies performed in Russia, and the experience with usage of mixed uranium-plutonium fuel in LWRs gained in the West, indicate a real possibility for using plutonium in the VVER-1000[4-5].

The simplest method for plutonium utilization in the VVER-1000 fuel cycle consists of direct replacement of a part of the uranium fuel by MOX fuel with no other significant changes required in the core design. It has been found that the use of MOX fuel results in altering the following neutronic characteristics of the core:

- Reduced (reactivity) worth of control rods and soluble boron;
- Reduced effective delayed neutron fraction;
- A less negative moderator temperature coefficient of reactivity at the end of cycle; and
- Increased power density peaking factors within the fuel assemblies.

At the same time, preliminary analyses of various designs for VVER-1000 cores with a 1/3 charge of MOX fuel assemblies revealed that the reactivity coefficients and peaking factors remain within the admissible design safety.

The question of introducing changes to the VVER-1000 power plant design (in particular, increasing the number of control rods) to facilitate use of plutonium fuel, was considered as part of the work on modernizing the VVER-1000 uranium fuel cycle. As noted above, in the modernized fuel cycle, the design safety margins during credible accidents is to be increased (for the VVER-1000 this increase amounts to about 20-25%). An additional increase in the control rod worth, also by 20-25% can be achieved by increasing the diameter of the absorber rods and by increasing the B^{10} content in the absorber rods. Hence, some of the reduction in the VVER-1000 design safety margins resulting from loadings of 1/3 core charges of plutonium fuel assemblies will be compensated by the above methods.

In Russia, design and manufacturing are being substantiated by experimental operation of three pilot MOX fuel assemblies in a VVER-1000 type reactor. Studies to substantiate the safety of VVER reactors containing 1/3 MOX fuel core loadings have commenced.

To ensure the high level of safety needed and to improve efficiency during the implementation of the modernized fuel cycles, the RRC "Kurchatov Institute", jointly with other enterprises, are carrying out intense studies in the directions described in the following sections:

TABLE 1. Main neutronic characteristics of modernized fuel cycles of VVER-440 (349 FA) (established refueling mode)

N	Characteristics	Note	Five-year fuel cycle		
			Variant I	Variant II	Variant III
1	Number of fuel assemblies loaded		78	78	78
2	Fuel assembly structural material		zirconium		
3	Reload fuel enrichment, % wt	average	4.28	4.12	4.28
		max for any fuel rod	4.40	4.40	4.60
4	Burnable poisons in reload fuel assemblies	absorber type	-	-	Cd_2O_3
5	Fuel cycle duration, effective full power days		313	307	322
6	Maximum power peaking factor during a cycle	over FA (Kq)	1.34	1.34	1.34
		over volume (Kv)	1.68	1.70	1.70
		over fuel rod in FA (Kk)	1.35	1.16	1.18
7	Maximum linear heat rate during a cycle (without allowance for the engineering hot channel factor), W/cm		260	241	242
8	Coolant temperature coefficient of reactivity, HZP, beginning of cycle, $1/°C*10^{-5}$	all control rods in the upper position	-5.8	-4.9	-6.3
9	Total control rod worth with the highest worth control rod out, minimum controlled power level, %	not lower	8.6	8.7	8.6
10	Temperature of recriticality at the end of cycle, all control rods fully in with one out , °C		58	38	62
11	Average burnup of discharge fuel, GWD/MTU		46	45	47
12	Specific consumption of natural uranium, kg/MW_t day		0.21	0.21	0.21

TABLE 2. Main neutronic characteristics of modernized fuel cycles for VVER-1000 units (established refueling mode)

N	Characteristic	Note	Fuel residence time	
			three year	four-year
1	Number of fuel assemblies loaded		54	42
2	Fuel assembly structural material		zirconium	
3	Reload fuel enrichment, % wt.	average	3.49	4.17
		maximum over fuel rod	3.70	4.40
4	Burnable poisons in reload fuel assemblies	absorber type	Gd_2O_3	Gd_2O_3
5	Fuel cycle duration, eff.day		294	291
6	Maximum power peaking factor during a cycle	over FA (Kq)	1.33	1.33
		over volume (Kv)	1.55	1.60
		over fuel rod (Kr)	1.47	1.47
7	Maximum linear heat rate during a cycle (without allowance for the engineering hot channel factor), W/cm		284	303
8	Coolant temperature coefficient of reactivity, HZP, beginning of cycle, $1/°C*10^{-5}$	all control rods in the upper position	-1	-1
9	Total control rod worth with the highest worth control rod out, minimum controlled power level, %	not lower	7.7	7.5
10	Temperature of recriticality at the end of cycle, all control rods fully in with one out , °C		80	149
11	Average burnup of unloaded fuel, GWD/MTU		37.5	47.7
12	Specific consumption of natural uranium, kg/MW_t day		0.21	0.20

4. Development of VVER Calculational Models and the Verification and Certification of Individual Computer Programs and Computer Program Systems

This development includes the following elements:

- A new version of the spectral analysis program used for the modeling of VVER reactor lattices is being worked out. The RRC KI library of evaluated data is used as input. The spectral analysis program is used to obtain broad group assembly average neutronic constants for fuel assemblies, taking into account a considerable degree of heterogeneity including the possible presence of Gd and Pu isotopes.
- The design and operation of a system of computer programs, offering the capability to perform detailed simulations at normal operating conditions for VVER reactors, are being improved. This computer code system is aimed at producing a capability to carry out simulations of modernized fuel cycles.
- The verification of the newly developed programs is being carried out for the purpose of their certification.

5. Provision of Computer Programs Ready for Operational Calculations to Nuclear Powerplants

A system of operational computer programs has been transferred to nuclear power plants in Russia and the Ukraine. At the RRC "Kurchatov Institute," specialists from nuclear power plants are being trained for participation in the possible use of the new computer code system directly at their nuclear power plants.

6. Development of Programs for Calculational Simulation of Transients and Emergency Situations at Nuclear Power Units

Two types of computer programs are being developed:

- A precise dynamic model (with allowance for feedback) of non-steady state neutron transport in VVER cores, taking into account possible motion of the control rods; and
- Dynamic modeling to describe transients in the VVER primary coolant circuit.

7. Development of Thermal-Hydraulic Calculational Programs

A specific feature of the current development work for the thermal-hydraulic analysis programs is to implement three-dimensional models to accurately take into account coolant mixing between individual fuel assemblies.

8. Experimental Reactor Physics Studies and Tests at Critical Facilities

At present, the RRC "Kurchatov Institute" has three critical facilities for testing VVER fuel. These are:

- The B-1000 full-scale critical facility used for tests of assembly configurations for uranium-water fuel lattices; and
- Two universal critical facilities intended for experiments to study neutron multiplying systems with various types of fuel assembly and fuel rod structures where, in particular, new types of fuel rods can be tested.

The test facilities can be used for measurements of 1) the spectral characteristics of fuel lattices, 2) spatial power distributions, 3) coefficients of reactivity, 4) control rod worths 5) the effects of thermal hydraulic conditions on neutronic characteristics and 6) the performance of in-core and ex-core diagnostics.

9. Generalization of Experience with Operating Nuclear Power Plant Reactors

The systematic acquisition and analysis of measured data characterizing the neutronic and thermal hydraulic characteristics of power plants in the process of startup and subsequent full power operation (power distribution, coefficients of reactivity, control rod worths, temperature, coolant flow rate and pressure) are being carried out. These investigations become particularly urgent at the time of initial operation with new fuel assembly designs as, for example, when the modernized fuel cycles are implemented.

10. Analysis of Fuel Rod Behavior and Reliability at High Fuel Burnups

The investigations are being carried out in four directions:
- Development of calculational models to describe changes to the thermal, physical and mechanical characteristics of the fuel at high burnups;
- Experimental studies of fuel rod behavior and properties at high fuel burnups;
- Detection of cladding failure in the fuel rods of assemblies removed from operating reactors at high burnups and the analysis of single cladding failures; and
- Analysis of fuel operational experience at operating reactors.

11. Upgrading of In-Core Instrumentation Systems

The speed, accuracy and reliability of the information obtained from in-core instrumentation is to be increased by correcting the signals from currently used rhodium detectors, by improving the temperature monitoring system and applying new computer programs for data analysis. In particular, the in-core instrumentation system's function will be improved by reducing the amount of neutron noise, by better organizing the information provided to the operator and by identifying anomalous modes at an early stage and mitigating their effects.

References

1. V. V. Saprykin, A. N. Novikov. *Characteristics of Four-year Fuel Cycles for VVER-440 Reactors with 349 and 313 Fuel Assemblies in Reactor Core.* Proceedings of the III Symposium of AER, 1993, 363-367.
2. V. Pavlov, A. Pavlovichev. *General Features of VVER-1000 Three and Four Batch 12-month Cycles with Improved Fuel Utilization.* Proceedings of the IV Symposium of AER, 1994, 575-582.
3. V. V. Saprykin, V. V. Sarbukov, V. I .Pavlov, A. M. Pavlovichev. *Experience of Utilization of New Design Fuel Assemblies in VVER-440 and VVER-1000.* Proceedings of the VI Symposium of AER, 1996, 613-626.
4. A. N. Novikov et al. *Use of MOX (R-Pu and W-Pu) Fuel in VVER-1000* NATO 1. Proc. of Meeting of Workshop on Managing the Plutonium Surplus: Applications and Options, London, 24-25 January, 1994.
5. N. N. Ponomarev-Stepnoy, E. S. Glushkov, I. K. Levina. *Burning of Weapon Grade Plutonium in VVER and HTGR Reactors.* NATO 3. Advanced Nuclear Systems Consuming Excess Plutonium, 203-217. 1997 Kluwer Academic Publishers.

USE OF URANIUM-ERBIUM AND PLUTONIUM-ERBIUM FUEL IN RBMK REACTORS

A. A. BALYGIN
G. B. DAVYDOVA
A. M. FEDOSOV
A. V. KRAYUSHKIN
Yu. A. TISHKIN
Russian Research Center "Kurchatov Institute"
Kurchatov Square
123182 Moscow
Russian Federation

A. I. KUPALOV-YAROPOLK
V. A. NIKOLAEV
RDIPE, a/b 788
101000 Moscow
Russian Federation

1. Introduction

The history of RBMK reactors is characterized by continuous transition from one type of fuel to another. The first RBMK units used 1.8% enriched fuel. After that, a decision was made to increase the enrichment to 2.0%. In 1986, it was decided to increase the enrichment further to 2.4%. Currently, transition to fuel with erbium poison and an enrichment of 2.6% is being implemented. These transitions were accomplished with relative ease due to on-line refuelling which is a special feature of RBMKs.

After the accident that occurred at Chernobyl Unit 4 in 1986, it was recognized that one of the main concerns in preventing future accidents would be reducing the large positive void reactivity coefficient of these reactors. Special measures were therefore undertaken to decrease the void reactivity coefficient in all RBMKs immediately. This goal was accomplished through the loading of about 80 additional neutron absorbers in each RBMK-1000. This measure was effective in decreasing the void reactivity coefficient, but it was not optimal from the point of view of fissile fuel economy. Investigations have been performed in recent years on how to achieve a more economical

T. A. Parish et al. (eds.),
Safety Issues Associated with Plutonium Involvement in the Nuclear Fuel Cycle, 121–130.
© 1999 *Kluwer Academic Publishers. Printed in the Netherlands.*

way of reaching the same level of safety. Many methods were considered and it was decided to implement a new type of fuel employing erbium burnable poison.

Due to its resonance in its absorption cross-section in the thermal energy range, erbium has a strong influence on the void reactivity coefficient. Also, thanks to the erbium poison, the uniformity of the power distribution may be improved. Some results from erbium fuel implementation in RBMKs are presented below. The question of using plutonium-erbium fuel in RBMKs is also discussed.

2. Description of Uranium-Erbium Fuel

Currently a new type of fuel -- uranium-erbium -- is being implemented in RBMKs. In this fuel, erbium (0.41% weight percent) is used as a burnable poison. The ^{167}Er isotope has a large resonance in its absorption cross section at 0.47 eV. Owing to this fact, erbium may have a strong influence on the void reactivity effect (VRE) because neutron absorption in erbium is increased when the coolant density is decreased.

Investigations have shown that the implementation of uranium-erbium fuel enhances reactor safety and fissile fuel economy. This is accomplished 1) by the unloading of absorbers added specifically to reduce the magnitude of the void reactivity coefficient, 2) by decreasing the average and maximum channel power, and 3) by increasing the burnup and decreasing the amount of spent fuel. In addition, Erbium has good compatibility with uranium dioxide. The strength and corrosion resistance of uranium-erbium fuels is the same as for normal uranium dioxide fuels. Calculations have shown that the most effective way to include erbium in the fuel was to add erbium oxide Er_2O_3, in all fuel pins of the assembly. Thermal conductivity and other fuel characteristics remain practically the same as for normal fuel due to the small amount of erbium used. Experiments performed with uranium-erbium fuel in research reactors has shown that radioactive gas release is slightly decreased when compared with normal fuel. The technology for mixing Er_2O_3 and uranium dioxide was developed at Mashinostroitelny Zavod in Elektrostal--the plant which currently produces fuel for RBMK reactors.

The substantiation of the possibility of using erbium as a burnable poison to decrease the void reactivity effect was performed on the basis of calculations using the STEPAN reactor code [1]. A two group cross section library was generated using WIMS-D4 [2]. It was also important in the beginning of the development of erbium fuel development to have an independent confirmation of the computational results. To verify the calculational procedure using STEPAN + WIMS-D4 special independent calculations were performed with the Monte-Carlo code MCNP [3]. The MCNP calculations used a library of cross sections obtained from ENDF/B-6 using NJOY which had point-wise energy dependence.

An infinite lattice was considered in the first step of the calculations. Dependences of the multiplication factor K_{inf} on burnup for the usual 2.4% enriched fuel and the 2.6% enriched fuel with erbium were obtained. The change of the void reactivity coefficient was calculated for a coolant density range from 0.5 to 0.0 g/cm^3. These results are presented in Fig. 1.

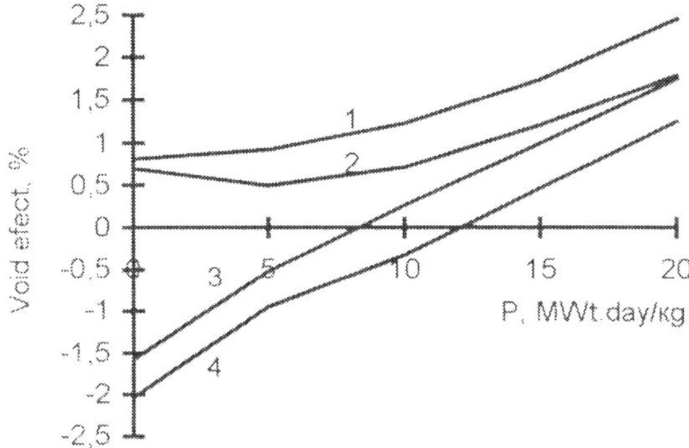

Fig. 1. Void reactivity effect (infinite lattice) for standard fuel (2.4%) using 1-WIMS, 2-MCNP and for uranium-erbium fuel (2.6%+0.41Er) using 3-WIMS and 4-MCNP

The MCNP calculations always gave larger lattice reactivity than WIMS-D4. The influence of erbium on the lattice reactivity and void reactivity effect is given in Table 1.

TABLE 1. Influence of erbium on lattice K_{inf} (ΔRO_{Er}) and void reactivity effect $\Delta(VRE)_{Er}$,%

Burnup, MWd/kg	ΔRO_{Er}, φ_{H2O}=0.5g/cm³		ΔRO_{Er}, φ_{H2O}=0g/cm³		$\Delta(VRE)_{Er}$	
	WIMS	MCNP	WIMS	MCNP	WIMS	MCNP
0.1	-9.86	-10.31	-11.98	-12.72	-2.36	-2.70
5.0	-4.50	-5.82	-5.86	-6.05	-1.43	-1.44
10.0	-1.19	-1.28	-2.13	-2.29	-0.96	-1.04
15.0	+0.83	+0.71	+0.01	+0.01	-0.75	-0.75
20.0	+2.03	+1.94	+1.32	+1.40	-0.70	-0.54

The presence of erbium decreases K_{inf} for small burnup levels while increasing it for large burnup levels. These effects are rather close in both the MCNP and WIMS-D4 calculations. WIMS-D4 shows a slightly smaller influence for erbium on the void reactivity effect at small burnups and a larger influence for erbium at large burnups (above 15 MWd/kg). The difference in the influence of erbium on void reactivity effect between MCNP and WIMS-D4 decreases with burnup.

The conclusion regarding the possibility of unloading additional absorbers due to the use of erbium fuel was based on results from the STEPAN+WIMS-D4 suite of codes. In order to prove this conclusion, calculations on a relatively complicated 12×12 3D polycell were performed using MCNP. The polycell approximates a real reactor at power reasonably well. The simplifications in geometry in the MCNP modeling were

124

minimal. Each fuel pin and each absorber was described in detail. Periodic boundary conditions were used on the sides of the polycell.

Two polycells were considered in order to model the Ignalina reactor's original condition (with standard 2% enriched fuel and with additional absorbers) and altered condition (with erbium fuel 2.4% enriched with 0.41 weight per cent of erbium and without additional absorbers). The control rod positions, axial distribution of the coolant density and the axial and radial burnup distributions were obtained from preliminary STEPAN calculations.

All of the fuel assemblies were grouped according to their burnup levels. Each variant was calculated with MCNP using $2 \cdot 10^6$ neutron histories. The statistical error in K_{eff} was about 0.08% and the error in the neutron fluxes was about 1 to 2%. The results from the void reactivity effect calculations are given in Table 2. It may be seen there that the MCNP and STEPAN-WIMS-D4 results are rather close.

TABLE 2. Void effect of fuel channels

	STEPAN		MCNP
	Continuous spectrum of burnup	Discrete spectrum of burnup	Discrete spectrum of burnup
Standard fuel, VRE, %	0.54	0.53	0.45±0.13
Erbium fuel, VRE, %	0.014	-0.08	-0.09±0.13
Change of VRE, %	-0.53	-0.61	-0.54

Full scale 3-D STEPAN code calculations were used to investigate the characteristics of RBMK-1000 and RBMK-1500 reactors with full loadings of erbium fuel and for the transition cores as the fuel was converted from normal to erbium fuel. According to these calculations, the weight percent of erbium was chosen to be 0.41 and the enrichment of uranium in the uranium-erbium fuel was selected to be 2.4% for RBMK-1500s and 2.6% for RBMK-1000s. The rates and sequences of unloading for the additional absorbers were also determined separately for each type of RBMK unit.

3. Operational Experience with Uranium-Erbium Fuel

It was decided to begin the actual transition with small portions of erbium fuel (150 - 200 FA (fuel assemblies)) without unloading AA (absorber assemblies). Predictive calculations were performed before the beginning of erbium fuel loading. It was decided to continue loadings of erbium fuel as the predictive calculations were confirmed. The first portion of 150 FAs was loaded into the Ignalina-2 unit in 1995. During the period from 1996-1997, the use of uranium - erbium fuel was expanded at that power station. The Ignalina-2 power plant presently has 658 FAs with erbium fuel and 24 AA have been unloaded as a result. Average burnup was increased by a factor of 1.15. The Ignalina-1 unit has been loaded with 150 FAs with erbium fuel and 8 AA have been unloaded. The first portion of 200 erbium FAs has been loaded in the Leningrad-2 unit. The operating experience with erbium fuel loadings have confirmed the accuracy of the predictive calculations. The first FAs are now half burned. Measurements of parameters related to safety, and especially the void reactivity coefficient, have been performed periodically during this loading. The measurements confirm the calculational predictions. The detailed characteristics of the erbium fuel loadings that have been carried out at various reactors are given in Tables 3-5.

TABLE 3. Characteristics of Ignalina unit-2 reactor (U-Er fuel loading)

Data	#* of U-Er FA	#** of AA	#*** of new control rods	Burnup**** MWd/Kg	VRC calculation	VRC experiment
26.06.95	0	53	0	851	0.81	0.83
25.07.95	25	53	0	840	0.76	0.70
08.08.95	50	53	0	846	0.70	0.70
15.08.95	74	53	0	834	0.59	0.62
08.12.95	100	53	24	851	0.62	0.66
05.12.95	125	53	24	836	0.52	0.56
29.01.96	150	53	24	842	0.46	0.53
25.06.96	195	53	24	848	0.49	0.6
30.06.95	195	49	24	859	0.64	0.6
16.07.96	236	49	24	858	0.53	0.6
23.07.96	240	45	24	868	0.67	0.7
22.11.96	275	45	48	903	0.59	-
11.12.96	300	41	48	919	0.70	0.7
27.02.96	397	33 (-20)	48	966 (13.5%)	0.68	0.75

*   Number of U-Er fuel assemblies (Fas) in the core
**   Number of additional absorbers (2477 rods) in the core
***   Number of control rods in new design (Aas) of the core (These control rods are implemented now in parallel with erbium fuel)
**** Average burnup

TABLE 4. Characteristics of Ignalina unit-1 reactor (U-Er fuel loading)

Data	# of U-Er FA	# of AA	Burnup MWd/Kg	VRE calculation	VRC experiment
23.12.96	0	53	846	0.95	1.0
21.01.97	100	53	814	0.59	0.7
23.01.97	100	49	818	0.73	0.7
07.02.97	140	49	812	0.57	0.7
14.02.97	140	45	827	0.63	0.7

TABLE 5. Characteristics of Leningrad unit-2 (loading of 200 assemblies U-Er fuel)

Date	# of U-Er FA	# of AA	# new control rods	Burnup MWd/Kg	VRC calc.	VRC, exp.
09.02.96	0	80	0	1333	0.78	0.7
11.05.96	25	80	0	1343	0.73	0.7
03.06.96	50	80	0	1338	0.68	0.7
11.07.96	101	80	0	1322	0.45	0.6
19.08.96	156	80	0	1310	0.25	0.6
15.10.96	190	80	16	1344	0.14	0.3
22.11.96	200	70	20	1374	-	0.3
25.01.97	200	60 (-20)	20	1429 (+7.2%)	-	0.5

4. Plutonium-Erbium Fuel

A technical report on the operational performance of new fuel assemblies with erbium is now in preparation. Additions to the Safety Analysis Report (TOB) for Ignalina are presently being written and are also being prepared for the Leningrad NPP.

The rather complicated problem of how to implement new uranium-erbium fuel in RBMK reactors has been solved. Another problem is how to use plutonium in RBMK reactors. RBMKs may be advantageous in the disposition of plutonium. It is well known that the amount of spent fuel from LWRs is increasing around the world. Plutonium from spent LWR fuel may, in principle, be used in RBMKs. Another source of plutonium is excess weapons-plutonium.

The most realistic way to dispose of plutonium is to use it as fuel in reactors which are presently operating. RBMK reactors are very suitable for this purpose because of their on-line refueling capability since this feature makes it relatively easy to introduce the loading of new fuel. Besides as of today, more than one half of the nuclear energy in Russia is produced by RBMKs.

The neutronic aspects of MOX fuel for use in RBMKs had been considered earlier [4]. The main challenges presented by such fuel are understood. The main difficulty is the large change in the multiplication factor and assembly power with burnup. This increases the non-uniformity of the power distribution in the reactor with on-line refueling. For plutonium fuel, the void reactivity effect remains positive similar to normal uranium fuel and neutron absorbers need to be added to keep the void reactivity coefficient negative.

The application of erbium-plutonium fuel represents a relatively new line of investigation into possible RBMK fuels. Some preliminary calculations of fuel variants based on uranium (0.3 weight percent enrichment)--plutonium--erbium have been made using WIMS-D4. Two plutonium isotopic contents were considered:
- plutonium discharged from VVER reactors; and
- weapons-grade plutonium.

The amount of plutonium and erbium in the fuel was varied such that the initial value of K_{inf} (for the fresh fuel) was not more than the K_{inf} for 2% enriched (ordinary) uranium fuel. A variant of the uranium-plutonium fuel without the addition of erbium was also considered for comparison purposes. In this case, the initial K_{inf} (for zero burnup) is equal to 1.25 for standard fuel (2% enriched uranium), and 1.26 for MOX with VVER plutonium and 1.30 for MOX with weapons-grade plutonium.

Due to fast plutonium burning, the rate of reactivity decrease with burnup is larger for fuel assemblies with MOX fuel as compared to standard fuel (Fig 2). The standard fuel is capable of reaching an average discharge burnup of 20 MWd/kg. The comparable discharge burnups for fuel based on VVER plutonium (2 % by weight) is 17 MWd/kg and for fuel based on weapons-grade plutonium (1.5 % by weight) is 17.5 MWd/kg. The initial powers of the fuel assemblies is larger for the plutonium fuel then for the standard fuel.

The addition of erbium with a simultaneous increase in the initial plutonium content significantly decreases the rate of reactivity decrease with burnup and leads to an increase in the discharge burnup without increasing the problems associated with the non-uniformity of the power distribution. The results of calculations for uranium-plutonium-erbium fuel are presented in Figs. 3 and 4.

128

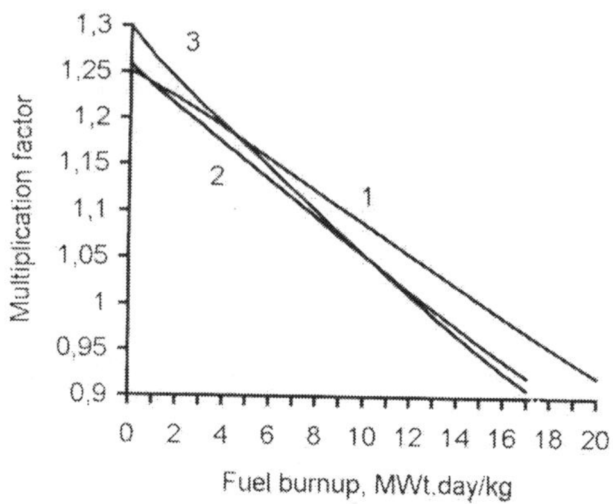

Fig. 2. Multiplication factor versus fuel burnup for standard and MOX fuel (1- 2% U235, 2- 2% Pu VVER, 3- 1.5% Pu weapons-grade).

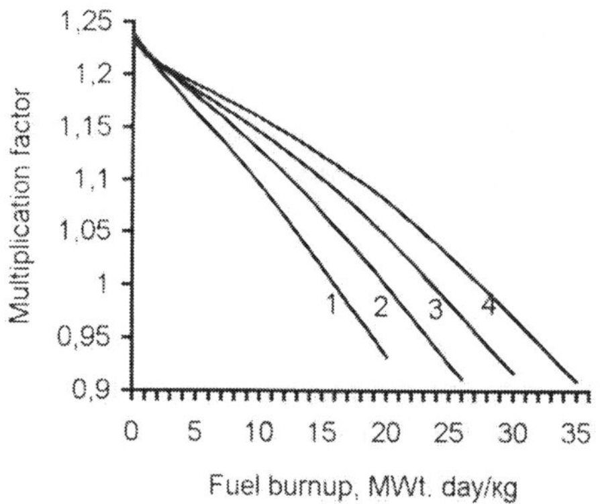

Fig. 3. Multiplication factor versus fuel burnup for uranium-plutonium-erbium fuel with weapon-grade plutonium.(1- 2% Pu+0.4%Er, 2- 2.5% Pu+0.6%Er, 3- 3% Pu+0.8%Er, 4- 3.5% Pu+1%Er)

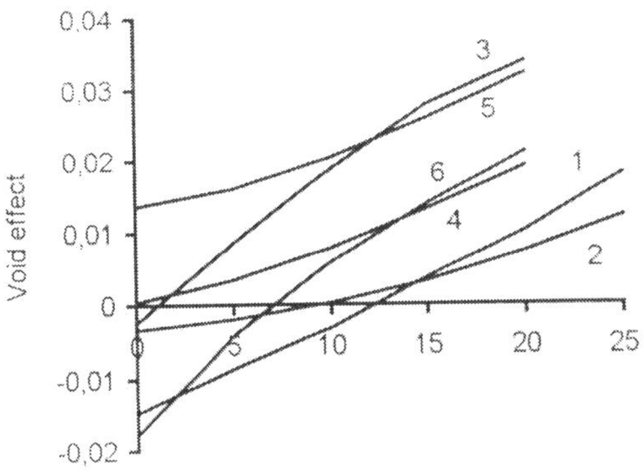

Fig. 4. Assembly void effect versus fuel burnup (1- 2.5% Pu weapons-grade+0.6%Er , 2-2.8% Pu VVER+0.4%Er, 3- 1.5% Pu weapons-grade, 4- 2% Pu VVER, 5- 2% standard, 6-2.4%+0.41% standard).

The main conclusion is that addition of erbium to MOX fuel allows for the achievement of a suitable value for the void reactivity coefficient and burnup levels while keeping non-uniformity of the power distribution to the level reached for uranium fuel.

5. Conclusions

In the last few years, a very important milestone has been passed in the field of RBMK fuel development. This consists of the implementation of uranium-erbium fuel. This implementation is currently being pursued in the cores of the Ignalina and Leningrad nuclear powerplants, and is planned for additional RBMKs. A large amount of calculational work has been performed to support the concept of erbium fuel. The results of calculations are in agreement with the characteristics observed at the reactors with erbium fuel.

Calculations indicate that uranium-plutonium-erbium fuel based on weapons-grade plutonium might also be successfully used in RBMKs. The addition of erbium burnable poison in such fuel opens the possibility for achieving improved burnup levels and operational characteristics for the core. RBMKs appear very suitable for MOX disposition due to their on-line refueling feature and their large experience base in the transitioning to new types of fuel.

References

1. Babaytsev M.N., Fedosov A.M., Glembotsky A.V., Krayushkin A.V., KubarevA.V., Romanenko V.S. (1993) *The STEPAN code for RBMK reactor calculation,* RRC "Kurchatov institute" IAE-5660/5.
2. Askew J.R., Fayers F.J., Kemshell P.B. (1996) *A General Description of the Lattice Code WIMS,* JBWES, Oct. p.564.
3. Briesmeister J.E. (1993) *MCNP4A - Monte-Carlo N-Particle Transport Code System,* Los Alamos National Laboratory report LA-12625-M.
4. Davydova G.B., Kvator V.M., Krayushkin A.V., Fedosov A.M. (1991) *The Improvement of RBMK fuel load,* Atomnaya Energia 70, 3-8.

PLUTONIUM MANAGEMENT AND ACTINIDE BURNING IN CANDU REACTORS

P. S. W. CHAN
M. J. N. GAGNON
P. G. BOCZAR
R. J. ELLIS
R. A. VERRALL
Atomic Energy of Canada Limited (AECL)
Sheridan Science and Technology Park
Mississauga, Ontario,
Canada L5K 1B2

1. Introduction

The fuel-cycle flexibility of CANDU reactors provides many options for the management of weapons-derived plutonium. These options include the use of conventional mixed-oxide (MOX) fuel for plutonium dispositioning as well as advanced options for plutonium annihilation. One of the advanced options, which uses an inert matrix (SiC) as the carrier for the weapons-derived plutonium, can also be used to destroy actinides created in civilian power reactors. Inherent safety features are incorporated into the design of the bundles carrying the plutonium and actinide fuels. This approach enables existing CANDU reactors to operate with various plutonium-based fuels, without requiring major changes to the current reactor design.

2. Conventional MOX Options for Disposition of Weapons-Derived Plutonium

Conventional MOX fuel presents a near-term, technically achievable, safe and economic option for disposition of weapons-derived plutonium. The main objective is not to destroy the plutonium, but to convert it to a form that has a high degree of diversion resistance with the characteristics of spent fuel, while producing electricity. The flexibility of the CANDU reactor allows the choice of several variants of this option, depending on the requirements and priorities. Important considerations are the timeliness of the deployment option, the plutonium disposition rate, and the economics. All of the

T. A. Parish et al. (eds.),
Safety Issues Associated with Plutonium Involvement in the Nuclear Fuel Cycle, 131–134.
© 1999 *Kluwer Academic Publishers. Printed in the Netherlands.*

MOX options use depleted uranium (0.2 wt % ^{235}U) as the base material in the fuel elements, which are arranged in 4 rings. Plutonium mixed with depleted uranium is used in the outer 2 fuel rings. The inner 2 fuel rings contain depleted uranium mixed with an appropriate amount of dysprosium, a burnable poison, designed to suppress the reactivity of the MOX fuel and to create a negative coolant-void reactivity of about -5.0 mk.

The first option, MOX(1), was chosen to provide the fastest start for the plutonium dispositioning mission. The reference fuel uses the standard 37-element geometry, with 37 fuel elements arranged in 4 rings of 1, 6, 12 and 18 elements, respectively. The inner 7 elements contain 5% dysprosium. The plutonium content in the elements of the third and fourth rings is 2.0% and 1.2%, respectively. This graded enrichment scheme minimizes the peak element power rating for a given bundle power by flattening the power distribution across the bundle. The fresh fuel contains 232 g of weapons-derived plutonium per bundle. The plutonium disposition rate in a Bruce A reactor is 1.0 Mg Pu per year (assuming an 80% capacity factor). The MOX fuel fabrication capacity requirement is about 80 Mg per year per reactor.

The second option, MOX(2), was optimized to maximize the plutonium disposition rate by increasing the amount of plutonium in the 37-element bundle without increasing the fuel burn-up significantly. To compensate for the excess reactivity, the burnable poison content in the central elements was increased from 5% to 15%, and the purity of the coolant and moderator was downgraded from 99.75% to 97%. The resultant fuel bundle has a plutonium loading of 300 g (3.1% Pu in ring 3, 1.6% Pu in ring 4), and 15% dysprosium in the central 7 elements. The plutonium disposition rate in a Bruce A reactor is 1.5 Mg Pu per year. The MOX fuel fabrication capacity requirement is 78 Mg per year per reactor.

The third option, MOX(3), increased the energy output that can be extracted from the weapons-derived plutonium. This could be an important factor if the plutonium is viewed as a valuable energy resource. This option uses a large amount of plutonium in the CANFLEX bundle, which has 43 fuel elements arranged in rings of 1, 7, 14 and 21 elements. The CANFLEX bundle has 20% lower peak element ratings than does the 37-element bundle operating at the same bundle power, and improved thermalhydraulic performance. The lower ratings facilitate achievement of higher burn-up, needed in this option, which has a core-average burn-up of 25 MWd/kg HE, about 3 times the nominal burn-up for natural-uranium fuel. A nominal D_2O purity of 99.75% is assumed. The plutonium fraction is 4.6% in ring 3 and 2.6% in ring 4. The central 8 elements contain 7% dysprosium mixed with depleted uranium. This advanced MOX bundle contains 472.6 g plutonium in the fresh fuel. The higher fuel burn-up reduces the MOX fuel fabrication requirement to 28 Mg per year per reactor, which would result in a significant reduction in the capital cost of the MOX fuel fabrication plant. The plutonium disposition rate is 0.8 Mg per year per reactor.

CANDU is a registered trademark of Atomic Energy of
Canada limited (AECL).

3. Advanced Options for Destruction of Weapons-Derived Plutonium

The CANDU system's versatility makes possible advanced options that can achieve near-complete destruction of the weapons-derived fissile plutonium. One such option is the use of an inert matrix (non-fertile material) as the carrier for weapons-derived plutonium. Another option, Pu-ThO$_2$ fuel, would also achieve a very high efficiency in plutonium destruction.

Some of the candidates under consideration for inert-matrix applications include SiC, MgAl$_2$O$_4$ (spinel), ZrSiO$_4$ (zircon), ZrO$_2$, CeO$_2$, CePO$_4$ and BeO. Based on detailed assessments of the mechanical, chemical and neutronic properties of these materials, AECL is focusing its efforts on SiC as the most promising candidate because of its very high thermal conductivity, its high melting temperature, and the absence of long-lived activation products resulting from irradiation.

The reference Pu-SiC fuel contains 250 g of weapons-derived plutonium mixed uniformly with SiC in the outer 30 elements of a standard 37-element bundle. The central element contains 20 g of gadolinium. Another 40 g of gadolinium is distributed uniformly over the 6 elements in the next ring. The distribution of the gadolinium, which has a very high depletion rate, was optimized to suppress the excess reactivity of the fresh fuel, to minimize the power ripple that is due to refueling and to give a negative coolant-void reactivity of -5.0 mk. About 94% of the fissile plutonium, or 77% of the total plutonium, is destroyed, and no new fissile material is created. Detailed fuel-management simulations with a full core of Pu-SiC fuel in a Bruce A reactor concluded that the reactor can operate within existing safety and operating envelopes.

The second CANDU plutonium-annihilation option uses ThO$_2$ as the carrier for the plutonium in a modified CANFLEX design in which the inner 8 elements are replaced with a large central graphite displacer. Plutonium at 2.6% (354 g per bundle) is mixed with thorium in the remaining 35 elements in the outer 2 fuel rings. This design gives a burn-up of 30 MWd/kg HE and a coolant-void reactivity of 8.6 mk. Ninety-four percent of the fissile plutonium, or 77% of the total plutonium is destroyed. About 168 g of [233]U is created per bundle and this useful energy resource can be recovered at a later time using a proliferation-resistant technology.

4. Actinide Burning in an Existing CANDU Reactor

The results for annihilating weapons-derived plutonium are directly applicable to the case of burning actinides in CANDU reactors. This was studied with a mixture of SiC and the actinides [237]Np, [241]Am, [243]Am and plutonium from spent PWR fuel in the elements of the standard 37-element bundle. A mixture of gadolinium and SiC was used in the inner 7 elements to suppress the excess reactivity of the fresh fuel. Several combinations of actinide inventory and gadolinium loading were considered. The best result was achieved using a full-core loading of actinide-SiC fuel containing 400 g of actinide (of which 356 g is Pu) and 60 g of gadolinium per bundle in a CANDU 6 reactor. The reactor performs within existing safety and licensing envelopes. About 60% of the total original actinide inventory is destroyed, as is 90% of the initial fissile plutonium inventory. The fueling rate is 9.2 bundles per full power day, resulting in the destruction of 0.68 Mg of actinides in a CANDU 6 reactor per year (assuming an 80% capacity factor).

5. Conclusion

The CANDU system provides unsurpassed flexibility for plutonium management through high neutron economy, on-line refueling, and a simple, economical fuel-bundle design. The CANDU MOX options offer timely and technically achievable ways of dispositioning weapons-derived plutonium. Advanced options, using an inert matrix or thorium fuel, can achieve more complete destruction of plutonium and actinides in existing CANDU reactors.

THE NEXT STAGE IN NUCLEAR POWER DEVELOPMENT: IMPROVED PU USAGE, SAFETY, RADIOACTIVE WASTE AND NON-PROLIFERATION FEATURES

V. V. ORLOV
Research and Development Institute of Power Engineering
Russia, Moscow
Fax:+095-975-20-19

The aims, status and prospects for nuclear power development are discussed. Large-scale nuclear power development will become both a reality and a socially acceptable way to solve global energy problems in the coming century if a new reactor type using a new fuel technology is adopted in the next few decades. This technological innovation must have the ability to fulfill the following requirements:

- Exploitation of the uranium-plutonium fuel cycle with breeding ratios near 1.0;
- Elimination of severe reactor accidents involving fuel failure and catastrophic radioactivity releases (inherent safety);
- Radioactive waste disposal at uranium mining sites without significantly increasing natural radiation levels;
- Reprocessing technology without uranium-plutonium separation, which enhances protection against the theft of potentially weapons usable fuel; and
- Economically competitive nuclear energy production by lowering nuclear power plant costs as compared to modern LWRs.

1. Motivation and Aims

The earth's population is predicted to double by the middle of the next century. This population increase is mainly expected in developing countries, and combined with bridging the gap in industrial development, is expected, at a minimum, to result in a doubling of worldwide primary energy consumption and in a trebling of worldwide electricity consumption.

135

T. A. Parish et al. (eds.),
Safety Issues Associated with Plutonium Involvement in the Nuclear Fuel Cycle, 135–138.
© 1999 *Kluwer Academic Publishers. Printed in the Netherlands.*

Utilization of energy from nuclear fission, based on a half-century of experience, is the most realistic way 1) to solve the energy problem, 2) to prevent international confrontation connected with exhaustion of inexpensive resources of hydrocarbon fuels and 3) to avoid the dangerous buildup of combustion products in the atmosphere. Nuclear power offers the capability to provide an important part (about one half) of the world's growing fuel needs if its capacity is increased by an order of magnitude by the middle of the next century and by two or three times more by the end of the next century. However, nuclear technology in its currently deployed (first stage), is not optimum for further large-scale deployment.

Based on predicted, inexpensive uranium resources ($\sim 10^7$ tons), thermal reactors consuming U-235 will not be able to exceed the currently achieved level of their contribution to world energy consumption (~ 5 %) and will only be able to satisfy the energy needs of a few countries and regions which have limited resources. In addition, first generation fast reactors have proven to be too expensive, and consequently, their construction has been restricted to a few demonstration units. Finally, issues related to nuclear power plant safety, radioactive waste management and non-proliferation of nuclear weapons within a closed fuel cycle may not have acceptable solutions with any of the currently deployed types of commercial power reactors.

Recently, a multitude of alternative reactor concepts have been proposed, but these studies have only been carried out at the level of scientific theoretical research. Therefore, it is difficult to anticipate a transition to technical development and demonstration for one of them without additional financial investment and adoption of a firm national policy. However, the experience already gained with nuclear power, and the investigations performed during recent years, have provided the information needed to support the choice of a specific reactor concept which is suitable for the next stage of nuclear power development and which is not very far from current engineering experience based on civil and military technologies.

The achievement of a closed fuel cycle, because its modern variants have been oriented toward extracting the Pu from spent fuel, is complicated by the fact that this approach could increase the risk of nuclear weapons proliferation. However, reactors having optimal safety and economic characteristics do not necessarily require extraction and/or addition of Pu. Such reactors may provide the opportunity to apply fuel reprocessing technology that precludes the possibility of its application to weapons-grade material production. The principal characteristics of potentially attractive reactor concepts are clear enough that a decision can be made on which type should be selected for the next stage of nuclear power development. Since this is the case, technical development of a reactor concept and construction of an experimental or demonstration unit may be accomplished within a time frame of 10 to 15 years

In the past, scientists were usually moved towards a particular reactor concept by the State which was interested in solving energy supply problems. The State also organized the design work and it was the reason for success. Nowadays, developing countries have the greatest interest in the large scale development of nuclear power and it is their initiative and their efforts to organize and to deploy reactors that can set the next stage of nuclear power development into motion. Undoubtedly, this initiative will be supported by Russian nuclear experts, and nuclear experts from other countries, who are seeking applications of their valuable engineering knowledge and experience. It is also logical for the developing countries to rely on governmental support from "nuclear" countries if new technology can decrease the risk of nuclear weapons proliferation. The challenge is to choose, develop and demonstrate a reactor and its associated closed fuel cycle which are able to satisfy the set of requirements dictated by the anticipated scale of energy consumption in the next century. These requirements lead to a rather definite choice for the reactor concept and fuel cycle technology.

2. Main Requirements for Future Nuclear Reactors and Fuel Cycle Technologies

The main requirements for the nuclear reactors and fuel cycle technology are as follows:

- At least, an order of magnitude reduction in natural uranium consumption, that is, accomplishment of a near breeding (BR~1) closed fuel cycle.
- Profitable and safe utilization of Pu which has been accumulated in the spent fuel from current nuclear power plants ($\sim 10^4$ tons) through its use in fast reactors using the U - Pu fuel cycle (with the initial Pu coming from spent fuel from current reactors and with first fuel loadings fabricated at plants of "nuclear" countries). It may be possible to organize international technological centers in such a way that these centers will be under regulation by "nuclear" countries.
- Elimination of the most serious reactor accidents (prompt supercriticality, loss of coolant, fire, steam and hydrogen explosions) that can cause fuel destruction and lead to large radioactive releases. This may be accomplished with the careful design an inherently safe fast reactor (BR~1) without a uranium blanket. Such a reactor would use dense thermoconductive nitride (UN - PuN) fuel, non-combustible, high temperature boiling lead (Pb) coolant, natural circulation to provide passive residual heat removal and strong negative temperature reactivity feedback.
- Production of naturally occurring radiation-equivalent radioactive wastes (for burial) by treatment of discharged fuel to remove actinides (decontamination factor of $\sim 10^{-3}$) for subsequent transmutation in burner reactors and by the extraction and utilization of Sr and Cs. The radioactive wastes will be put in a stable physical and chemical form to impair migration from storage/burial sites.
- Utilization of fuel reprocessing technology which eliminates the possibility of Pu extraction, and hence, reduces nuclear weapons proliferation risks. The fuel reprocessing technology must be appropriate for use with a fast reactor with equilibrium fuel composition (BR~1) and without a uranium blanket. This will involve further research and development of a fuel reprocessing technology which depends mainly on fission product extraction (atomic mass separation).
- Physical protection of fuel from theft. Quasi-continuous refueling which eliminates outside of reactor fuel storage is called for with the fuel

processing facility located near the nuclear power plant. The extraction of Pu from the spent fuel of thermal reactors which has been stored in cooling ponds must be accomplished under highly protected conditions during both fuel reprocessing/ fabrication and usage at the reactors.

- Reduction of the electricity production cost. The cost of electricity from a nuclear power plant   using a naturally safe fast reactor may be lower than the cost of modern nuclear power plants like LWRs because of the high efficiency of fuel (level of fuel burnup above 10% of HM) usage and thermal energy conversion (thermal efficiency above 40%).

3. Conclusions

The nuclear reactor concept chosen for future large scale nuclear power development should be based on demonstrated technologies and must use well known materials in order to limit the time and cost necessary for development and deployment. Lead (Pb) has physical and chemical properties which are very similar to those of Pb-Bi, which has been used for naval reactors in Russia, and hence, for which there is a large engineering experience base. Only the melting temperature of Pb is different from melting temperature of Pb-Bi. Experiments, which have been performed with Pb coolant loops, have proven the applicability of the construction materials and anticorrosion technology used for Pb-Bi to Pb systems.

The choices of operational temperatures and cooling system characteristics need to provide for the avoidance of Pb freezing during accidents. The properties of UN-PuN fuel as a function of burnup have been studied in many reactor experiments. Detailed experimental investigations of Pb cooling during fuel irradiation still remain to be performed at anticipated reactor conditions.

NEUTRONIC ASPECTS OF WEAPONS-GRADE PLUTONIUM UTILIZATION IN THE VVER-1000 FUEL CYCLE

N.I. BELOUSOV
V.I. NAUMOV
V.I. SAVANDER
Moscow State Engineering Physics Institute (Technical University)
Department of Theoretical and Experimental Reactor Physics
Kashirskoe shosse, 31
Russia 115409, Moscow
Russian Federation

1. Introduction

The utilization of weapons grade plutonium as part of the reload fuel for operating advanced light water reactors is one of the most realistic ways of disposing of excess plutonium using nuclear energy technology. It can be concluded, on the basis of European experience with MOX fuel for light water reactors and on the basis of published results from investigations which were performed for Russian VVER-1000 reactors, that a 30 % loading of weapons-grade plutonium, ie., 30% of the total amount of fissile material in the core is made up of weapons grade plutonium, will provide operational performance which is close to the operational performance of a reactor with a conventional uranium loading [1,2].

With the possible objective of increasing the amount of plutonium which can be used to fuel modified existing reactors or new types of reactors, it would seem to be interesting to perform an investigation to determine the operational characteristics of light water reactors fueled with plutonium in order to specify their safety performance and control system requirements, and to compare these with their values for conventional uranium fueled versions. In this paper, an investigation of uranium - plutonium fueled light-water reactor systems has been performed based on the standard parameters for Russian VVER-1000 fuel assemblies. It should be noted that the assumed usage of the standard VVER-1000 assembly configuration does not eliminate the need for further investigations of uranium - plutonium light water assembly designs in order to obtain optimized lattices which may have more attractive safety characteristics and may achieve more efficient plutonium utilization.

139

T. A. Parish et al. (eds.),
Safety Issues Associated with Plutonium Involvement in the Nuclear Fuel Cycle, 139–146.
© 1999 *Kluwer Academic Publishers. Printed in the Netherlands.*

If the question of weapons-grade plutonium disposition is considered for the case of operating VVER-1000 reactors, the main problems consist of determining the fuel loading and plutonium burning capability of the reactor core. The main factors limiting plutonium disposition from a practical engineering standpoint will be, firstly, conserving the currently achieved safety margins, and, secondly, minimizing the modifications to the standard refueling scheme and safety systems of the reactor. As a result, if utilization of plutonium is to be undertaken, these two main questions need to be answered quantitatively. The first step in answering these questions consists of performing an analysis to obtain the plutonium loading within a fuel assembly and optimal assembly loading patterns in order to obtain the target cycle energy production that meets power peaking constraints. [1,2]. The second step consists of an analysis of reactor operational performance and to determine safety margins for the case of mixed plutonium-uranium fuel loadings.

2. Statement of the Problem

Comparative analysis of the neutronic and dynamic characteristics of VVER-1000 reactor cores with uranium and plutonium fuel loadings have been performed and the results are presented in this paper. The calculational investigations were carried out on individual cell and polycell levels without detailed consideration of the reactor core layout because the main goal of this study was only to investigate the fundamental physics features of plutonium utilization in VVER-1000 reactors. Within the present study, all of the calculations were carried out using the GETERA computer code developed at MEPhI. This code is widely used for such types of investigations in Russia [3].

A VVER-1000 fuel assembly with a U-235 enrichment of 4.4 % was considered as the reference design for the investigations. The neutronic characteristics of the fuel assembly were modeled on the level of an elementary unit cell which consists of three geometrical zones. The configuration and composition of the unit cell considered are shown in Fig. 1.

The outer zone of the unit cell contains a homogeneous mixture of water and steel to model the fuel rod spacers and other structural components of the fuel assembly. This zone of the unit cell is considered as the moderator zone. The reactor core was represented with a polycell model which consists of three types of fuel assemblies with different enrichments in order to model a three batch refuelling scheme (each batch resides for three cycles).

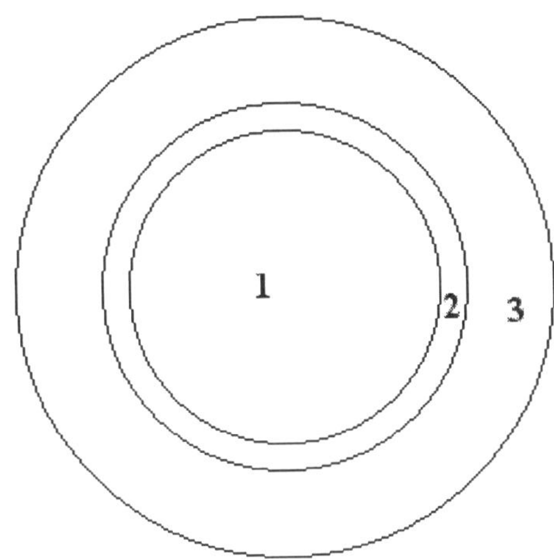

Fig. 1. Three-region VVER cell.

1. FUEL UO_2 $\gamma=10g/cm^3$ $R_1 = 0.388$ cm
2. CLADDING   Zr $R_2 = 0.455$ cm
3. MODERATOR H_2O $\gamma=0.71g/cm^3$ $R_3 = 0.669$ cm

The relationships between assemblies with different levels of burnup were determined using an albedo matrix representation of neutron currents. The elements of the albedo matrix are equal to the fraction of the neutron current from cells of one type to cells of another type. Excess reactivity is compensated by soluble boron, which is included in the moderator zone. Calculations of neutronic characteristics were performed for two different reactor states: 1) beginning of cycle and 2) end of cycle with a cycle length of one year.

In the beginning of cycle state, the polycell (core) consists of one third fresh fuel, one third fuel irradiated for one cycle and one third fuel irradiated for two cycles. In the end of cycle state, the polycell is composed of equal amounts of fuel irradiated for one, two and three cycles, respectively.

At the beginning of cycle, the core is in its most reactive (neutronically) state with a core average burnup equal to the cycle burnup. The excess reactivity in the beginning of cycle state is compensated by the addition of soluble boron to the reactor water. At the end of cycle, the core is in its least reactive state with a core average burnup equal to twice the cycle burnup. Also, at the end of cycle, there is no longer any excess reactivity and the soluble boron content of the reactor water is zero. Schematic descriptions of these core states are given in Fig. 2.

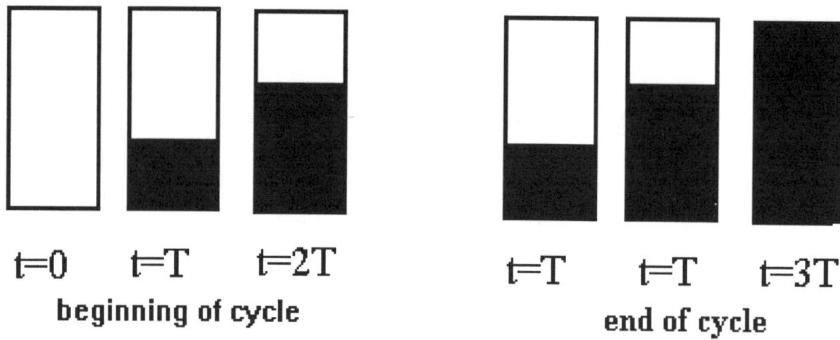

Fig. 2. Configuration of polycells for conditions at the beginning and at the end of an irradiation cycle.

3. Brief Description of the GETERA Program Code

The GETERA computer code has the capability to calculate the space-dependent power distribution in multi-region heterogeneous polycells for plane, cylindrical and spherical geometries using the multigroup approximation and the method of first collision probabilities.

Neutronic calculations in the slowing-down energy range (10.5 Mev - 2.15 ev) are made on the basis of the BNAB-80 cross section library with 22 energy groups. The effects of resonance shielding are taken into account using equivalence relations during the calculation of the group constants for the resonance nuclides. Thermal spectra calculations are made on the basis of the differential Kadillak with the utilization of microconstants obtained from an evaluated nuclear data file (ENDF/B - IV).

4. Main Results of the Calculational Analysis

If plutonium and uranium are utilized as fuels in thermal reactors, the main differences from a neutronic point of view can be summarized as follows:

- The absolute values of the microscopic cross sections for absorption and fission of plutonium are two times higher than those of uranium;
- The cross section of Pu-239 has a large low energy resonance at E = 0.3 ev; and
- The value of the effective delayed neutron fraction for plutonium is much less than that for uranium.

TABLE 1. Excess reactivity behavior as function of loaded plutonium fraction

VALUE	CASE 1 0.%WGPu	CASE 2 1/3WGPu	CASE 3 2/3WGPu	CASE 4 100%WGPu
Initial Excess Reactivity	0,093	0,085	0,080	0,078
Initial Concentration of Boron Absorber, %	100	127	145	161
Relative Worth of Control Rod System (beginning of cycle), %	100	70		56
Relative Worth of Steady-state Xe and Sm Poisons (beginning of cycle), %	100	86	77	68
Relative Worth of Control Rod System (end of cycle), %	114			76
Relative Worth of Steady-state Xe and Sm Poisons (end of cycle), %	175	186	149	140
Relative Change in Effective Delayed Neutron Fraction , %	100	76	62	52

Preservation of the conventional fuel cycle parameters which are typical of a uranium fueled core, such as a discharge burnup level of 40.0 MWD/MTU and a cycle length of 300 effective (full power) days, were the main conditions imposed when evaluating the effects of specific loadings of plutonium.

The number of neutrons produced per fission for Pu-239 is higher than that for U-235, but the infinite neutron multiplication factor is practically unchanged for the same enrichment in a pure uranium-water lattice and in a pure plutonium-water lattice because the ratio of capture-to-fission for plutonium is much higher in the thermal energy range. As a result, it is possible to achieve the same level of burnup for plutonium fuel by using a plutonium loading equal to the enrichment of uranium (4.4 %).

The relative amount of thermal neutron absorption is increased in the MOX fuel, and this in turn reduces the efficiency of the neutron absorbers used for reactivity compensation (soluble boron, control rods, and burnable absorber). This situation means that plutonium fueled cores will require a higher soluble boron concentration in the beginning of cycle state, as is shown in Table 1.

The reduction of the control system's efficiency is also considerable and the control rod system worth is reduced to 56 % of the worth for a uranium core at a plutonium loading of 100%. The effective delayed neutron fraction is also reduced considerably from its value for the reference case of a uranium core as the plutonium loading is increased.

144

The most important dynamic parameters, which determine the safety margins during normal operation, are summarized below:

$$\alpha_f = \frac{\partial \rho}{\partial T_f}$$ - the fuel temperature coefficient of reactivity;

α_m - the moderator temperature coefficient of reactivity (constant pressure)

The latter coefficient, α_m, can be represented as a sum of two different components as follows:

$$\alpha_m = \frac{\partial \rho}{\partial T_m} + \frac{\partial \rho}{\partial \gamma_m} * \frac{\partial \gamma_m}{\partial T_m} \qquad (1)$$

using the partial moderator temperature coefficient of reactivity at constant mass density of the moderator ($\frac{\partial \rho}{\partial T_m}$) and the partial mass density coefficient of reactivity at constant moderator temperature ($\frac{\partial \rho}{\partial \gamma_m}$).

TABLE 2. Coefficients of reactivity as a function of Pu loading (beginning of cycle)

VALUE	CASE 1	CASE 2	CASE 3	CASE 4
$\alpha_f * 10^5$, 1/°C	- 1.95	-1.93	-1.95	-1.96
$\frac{\partial \rho}{\partial T_m}$, 1/°C	2.52	3.83	3.66	3.13
$\frac{\partial \rho}{\partial \gamma_m}$, cm3/g	0.076	0.108	0.125	0.140
$\alpha_m * 10^5$, 1/°C	-13.5	-18.8	-22.7	-26.3

Calculational results for the temperature coefficients of reactivity for uranium and plutonium cores are shown in Table 2. The data presented there demonstrate that the reduction of the coolant's importance as a neutron absorber in the lattice when plutonium fuel is loaded increases the mass density coefficient of reactivity at constant moderator temperature. As a result, the moderator temperature coefficient of reactivity has a negative value for the full range of plutonium loadings. However, the just noted behavior of the moderator temperature coefficient of reactivity will provide considerable excess (positive) reactivity for accidents that inject cold water into the reactor core. For example, the range of temperature reduction, which will give an excess (positive) reactivity equal to the reactivity worth of the effective delayed neutron fraction, is four times lower for a 100% plutonium loading than for the uranium reference case. This behavior is an important limitation for the large-scale utilization of plutonium in the fuel cycle of existing VVER-1000 reactors. On the other hand, it needs to be considered that the same amount of excess (positive) reactivity can be compensated for by fuel temperature increases which are two times lower in the plutonium variants than in uranium reference case.

5. Conclusions

The results presented here support the conclusion that utilization of plutonium, which has considerably higher cross sections than U-235 for capture and fission in the neutron spectra characteristic of uranium-water lattices, reduces the reactivity worth of the control system and soluble boron absorber.

In addition, plutonium utilization in VVER-1000 reactors leads to a considerably lower value for the effective delayed neutron fraction, and hence, increases the sensitivity of the reactor's power level to changes in operational parameters (fuel temperature, moderator temperature and density) that affect the value of the effective neutron multiplication factor. Attention to these issues needs to be emphasized in designing VVER-1000 cores to dispose of plutonium. It therefore appears that the development of systems for large scale, efficient plutonium disposition will probably require the design of new reactors which are less sensitive to operational parameter changes and which have a higher level of control of these parameters.

Because weapons-grade plutonium has low radioactivity, this material can be used in order to develop and demonstrate the fuel fabrication technology for plutonium usage within a closed light water fuel cycle. Therefore, it is also important to perform investigations of how to efficiently utilize plutonium in advanced light-water reactors with high levels of burnup. Efficient plutonium utilization involves increasing the level of burnup, reducing the number of fueling operations, decreasing the amounts of spent fuel to be stored, and improving the neutronic characteristics and safety margins.

More efficient plutonium utilization can be achieved if the startup loading can be increased to a higher plutonium fraction through an optimal choice of water to fuel fraction and utilization of different burnable absorbers.

References

1. Novikov A.N., and al. (1994). Use of MOX (Rpu and Wpu) fuel in VVER-1000 (neutron - physical aspects of possibilities), *Proceedings of the NATO Advanced*

Research Workshop on Managing the Plutonium Surplus: Applications and Technical Options, London, 93 - 96.

2. Levina I.K., Saprykin V.V., Morozov A.G. (1994) The safety criteria and VVER core modification for weapon plutonium utilization, *Proceedings of the NATO Advanced Research Workshop on Mixed Oxide fuel (MOX) Expluatation and Destruction in Power Reactors, Obninsk, Russia, 83 - 92.*

3. Belousov N.I., and al. (1992) The code GETERA for cell and polycell calculations: models and capabilities, *Proceedings of the 1992 Topical Meeting on Advances in Reactor Physics, 2, 516.*

VALIDATION OF THE MCU-RFFI/A CODE FOR APPLICATIONS TO PLUTONIUM SYSTEMS AND USE OF THE MCU-RFFI/A CODE FOR VERIFICATION OF PHYSICS DESIGN CODES INTENDED FOR CALCULATIONS OF VVER REACTOR PERFORMANCE WITH MOX FUEL

M. A. KALUGIN
Russian Research Center "Kurchatov Institute"
Kurchatov Square 1, Moscow, 123182, Russia

Abstract

MCU-RFFI/A is a general-purpose Monte Carlo code for solving neutron transport problems in the energy range between 20 MeV and 10^{-5} eV. The main applications of the MCU-RFFI/A include nuclear safety assessments, nuclear data validation, design calculations, verification of design codes. Validation of MCU-RFFI/A for applications to plutonium systems and verification of physics codes for VVER reactors using MOX fuel are described.

1. The MCU-RFFI/A Monte Carlo Code and its Library DLC/MCUDAT-1.0

MCU-RFFI/A is a continuous energy Monte Carlo code. It is used with both pointwise and step function representations of the energy dependence of neutron cross sections. To describe cross sections in the unresolved resonance range, the subgroup method (similar to the probability table method) is used. It is also possible to use a detailed description of the energy dependence of cross sections in the resolved resonance range. For the most important isotopes, an "infinite" number of energy points is used to describe the variation of the resonance cross sections. In this case, cross sections are calculated during the Monte Carlo run at every energy point on the basis of analytical formulations and a resonance parameter library. This permits one to perform calculations without preliminary tabulation of resonance cross sections and allows the user to estimate temperature effects independently of the cross section library state. Special fast algorithms to calculate resonance cross sections are used. For the thermal energy range, Monte Carlo sampling is performed using $S(\alpha,\beta)$ scattering laws (and coherent elastic

147

T. A. Parish et al. (eds.),
Safety Issues Associated with Plutonium Involvement in the Nuclear Fuel Cycle, 147–158.
© 1999 *Kluwer Academic Publishers. Printed in the Netherlands.*

scattering cross sections) or a free gas model. One may solve problems taking into account both prompt neutron and delayed neutron energy spectra.

The accuracy of modern Monte Carlo codes is limited mainly by the accuracy of the nuclear data used. The verification and validation of the MCU-RFFI/A code was carried out using the nuclear data in the DLC/MCUDAT-1.0 data library. This library includes: the ACE library prepared using the NJOY code with ENDF/B-VI evaluated data;

- the ABBN 26 group library;
- the LIPAR library of nuclear resonance parameters for the resolved resonance region;
- the VESTA library of continuous energy thermal neutron cross sections;
- the TEPCON thermal neutron data multigroup library which is an alternative to VESTA; it contains cross sections for energies less than 1 eV at various temperatures and includes the zero and the first angular moments of the scattering cross sections.

The scattering laws for the VESTA and TEPCON libraries are calculated using the ENDF phonon spectra set.

While the ACE library may be used down to the boundary of the thermal energy region in criticality calculations, it should, as a rule, be used only in the fast region above 100 KeV. Alternatively, the ABBN library is used to describe smooth cross sections (via step functions) for energies above the resolved resonance region during a continuous energy Monte Carlo run. Inelastic scattering is modeled using a probability density step functions. The LIPAR data are used directly in the Monte Carlo run for calculating the continuous energy cross sections, while taking into account Doppler broadening. Both the continuous energy and the multigroup approximations can be applied in the thermalization energy range. The VESTA library is used in the first instance, and the TEPCON library in the latter case. For the most important moderators, the scattering cross sections are calculated using Gauss's incoherent approximation. The elastic coherent cross sections are used if necessary. DLC/MCUDAT-1.0 also includes the DOSIM library which contains data for dosimetry reaction rate calculations. The DOSIM library is prepared by the NJOY code which uses IRDF, ENDF/B-VI and other files for input. The DLC/MCUDAT-1.0 library contains information for 131 isotopes.

MCU-RFFI/A allows the user to model arbitrary three-dimensional systems described by combinatorial geometry. The geometry zones are represented by Boolean combinations of 13 different bodies. One may describe lattices with repeated elements by defining the repeated elements and translation vectors. Lattices may also include nonuniformities. These are given in the applications presented later. For zones and repeated elements, the following attributes are defined by the user: tally zone number, tally object number, material number, etc. Any number of geometric zones may be united into one tally zone or tally object. For lattices, these attributes may be generated automatically.

Boundary conditions include vacuum, white and specular reflection, the condition of translational symmetry and symmetry planes. In criticality calculations, neutron leakage given by a buckling vector may be calculated by solving Benoist's problem. Double heterogeneous systems with fuel elements containing many thousands of spherical microcells can also be modeled.

By default, MCU-RFFI/A calculates a set of flux functionals integrated over the entire system, over the tally zones or over tally objects. Energy bins are specified by the user. The following parameters are calculated: effective multiplication factor, migration lengths, neutron lifetime, generation time, effective fraction of delayed neutrons, fluxes and reaction rates for isotopes in the tally zones, the few group constants for tally objects, and spectral indexes. The user may also calculate the currents, the surface fluxes and other surface flux functionals for the edges of the tally objects. MCU allows one to determine the diffusion coefficients of reactor cells depending on different definitions, on the cell position, and on the edge number. For Benoist's problem, the multiplication factor's dependence on buckling is calculated. The dependence of the effective multiplication factor on buckling may be determined during a single code run and the critical buckling may be determined. The solution of this problem is used to calculate Benoist and Bonalumi-type diffusion coefficients. The variances of all of the functionals are calculated including fraction functionals similar to spectral indexes or average cross sections.

External source distributions may be specified using functions defined in the different geometry zones, and may be quite general. Standard variance reduction techniques are used, like splitting, roulette, weight window, and various combined estimators for the multiplication factor. The new ALIGR technique [1] was implemented to determine the reaction rates in very small monitors located in the experimental channels of large research reactors. It seems that the ALIGR technique should also be applicable to other difficult problems, including shielding problems. The MCU-RFFI/A code modules are written in FORTRAN 77. The full description of the code and its nuclear data library is available in Russian. An MCU-RFFI/A user's guide has been translated into English.

MCU-RFFI/A is a part of the MCU project. This project includes several tasks such as developing nuclear data libraries as well as the codes for processing these libraries. Other tasks include designing and updating the Monte Carlo code package, and supporting and updating the CLAD library containing the MCU input lists and the results of MCU calculations for various problems. The MCU-RFFI/A code is assembled from modules of the MCU code package. The modular code architecture allows the user to solve transport problems using a set of various algorithms and mutually-changeable program modules, while also utilizing different data libraries. The MCU package consists of a set of segments and each segment has its functional destination and interfaces strictly determined by the code architecture. The MCU module types are classified as follows: C=Control, P=Physical, G=Geometric, T=Tallies, S=Sources. As a rule, every type of module has alternative versions corresponding to specific modeling algorithms. The executable program variant (EP) is assembled according to the equation:

$$EP=C+P+G+T+S \qquad (1)$$

where C,P,G,T,S are the modules described above. The package architecture allows one to assemble the code in order to simulate random walks for several types of particles. The module interfaces are very simple and allow a user to include the modules in different codes or to change the memory allocation between the modules in the framework of one code. The TERMAC and STEN codes developed within the MCU project are used for calculating scattering laws and the thermal neutron elastic coherent cross sections (VESTA and TEPCON libraries) by utilizing the phonon spectrum library. GRAF is a two-dimensional color graphics, solid geometry modeler. It permits one to check initial geometry data and look at any plane section of the system being considered. The distribution of materials, geometry zones, tally zones and tally objects may be seen.

The MCU project has been carried out at the Kurchatov Institute beginning in 1982. Four freeware MCU code versions have been issued thus far. These are MCU-1 (1985), MCU-2 (1990), MCU-3 (1993) and MCU-RFFI (1995). MCU has been validated using data from about 300 experimental assemblies (including all known issued benchmarks before 1995) and has also been verified by solving thousands of practical problems. MCU is widely used in Russia for calculating parameters for practically all types of Russian reactors and calculations for various non-reactor applications, like in-pool storage, shipping casks etc. The CLAD library contains descriptions of more than 300 different experimental assemblies.

The authors of the MCU-RFFI/A code are: Abagyan L.P., Alexeyev N.I., Bryzgalov V.I., Veretenov V.V., Glushkov A.E., Gomin E.A., Gurevich M.I., Kalugin M.A., Maiorov L.V., Marin S.V., and Yudkevich M.S. In 1996 the MCU-RFFI/A code with its library DLC/MCUDAT-1.0 were certified by the Russian Gosatomnadzor as a tool to perform criticality calculations on practically any type of neutron multiplying system including ones with plutonium and mixed (uranium-plutonium) fuel. The MCU-RFFI/A code is distributed on a commercial basis.

2. Comparison of the MCU-RFFI/A Calculations with Experimental Data and Other Code Results in Criticality Safety

Validation of the MCU-RFFI/A code with respect to the safety of uranium-plutonium systems was accomplished by comparison of the calculated results to critical benchmark experiments [2]. The high reliability of the code and its neutron data library has been demonstrated for a wide range of neutron energy spectra in systems with plutonium.

During reprocessing of irradiated fuel rods and during plutonium fuel fabrication, water solutions containing salts of plutonium, frequently in a mix with salts of uranium, are widely used. Detailed attention is given to the accuracy of the computational prediction of criticality for such systems. Extensive experimental data, enabling computational codes and their neutron cross section data to be tested, have been published in [3].

Results of K_{eff} calculations for critical assemblies performed using the MCU-RFFI/A code with the DLC/MCUDAT-1.0 library simulating typical situations that appear during operations involving plutonium are presented in the next section. The description of most of the assemblies and computational results using the MCNP code with ENDF/B-V library are taken from the already mentioned collection [3]. Only a very general description of the experimental configurations and structure is given here. In all of the tables presenting computational results, the values in parentheses, following the calculated K_{eff}, indicate the standard deviations due to statistics.

2.1 WATER-REFLECTED 9, 10, 11, AND 12 INCH DIAMETER CYLINDERS OF PLUTONIUM NITRATE SOLUTIONS   (NEA/NCS/DOC(95)03/I PU-SOL-THERM-010)

This set of experiments consists of thin-walled vessels containing a $Pu(NO_3)_2$ solution that are surrounded by an infinite (in a practical sense) reflector of water. The content of ^{240}Pu is about 3 % . The assemblies differ according to the diameters of the cylinders and the concentrations of plutonium they contain. Criticality is achieved by varying the level of the solution in the cylinder. The experimental error for the effective neutron multiplication factor is ±0.0048.

TABLE 2.1. Critical cylinders containing a plutonium nitrate solution

Case	D	Pu, g/l	H/ ^{239}Pu	K_{eff}	
	(inch)			MCNP	MCU
9-1	9	99.09	266.9	1.0212(10)	1.0137(13)
9-2	"	73.92	356.9	1.0197(10)	1.0096(13)
9-3	"	54.53	484.2	1.0153(10)	1.0055(13)
11-1	11	54.43	485.0	1.0189(10)	1.0106(11)
11-2	"	47.21	558.1	1.0168(10)	1.0080(9)
11-3	"	47.21	558.1	1.0183(10)	1.0079(13)
11-4	"	41.73	605.9	1.0076(10)	1.0001(9)
11-5	"	36.90	665.4	1.0097(9)	1.0027(9)
11-6	"	63.99	414.3	1.0205(10)	1.0121(9)
11-7	"	48.98	535.2	1.0087(10)	0.9985(13)
12-1	12	48.75	543.4	1.0163(10)	1.0078(11)
12-2	"	42.29	618.3	1.0070(10)	1.0070(8)
12-3	"	36.52	728.1	1.0022(9)	1.0132(9)
12-4	"	31.14	849.7	1.0165(9)	1.0086(13)
Average K_{eff} value over all the variants				1.0142±0.0060	1.0075±0.0046

2.2 WATER-REFLECTED SPHERES OF PLUTONIUM NITRATE SOLUTIONS (NEA/NCS/DOC(95)03/IB, PU-SOL-THERM-01÷03)

This set of experiments consists of thin-walled spherical vessels containing a $Pu(NO_3)_2$ solution. Each sphere is surrounded by an infinite (in a practical sense) reflector of water.

The ^{240}Pu contents are 2, 3 or 4.5 %. The assemblies differ according to the sphere diameters and the plutonium concentrations.

TABLE 2.2. Critical assembly with a plutonium nitrate solution

Assembly/	D	Pu	H/ ^{239}Pu	K_{eff}		
Variant	inch	g/l		experiment	MCNP	MCU
PU-...-001 1	11.5	73.0	371	1.0000(50)	1.0117(10)	0.9995(13)
PU-...-001 5	"	140.0	180	1.0000(50)	1.0100(10)	1.0037(11)
PU-...-001 6	"	268.7	91	1.0000(50)	1.0144(10)	1.0029(13)
PU-...-002 1	12.0	49.84	524	1.0000(47)	1.0126(10)	1.0007(12)
PU-...-002 4	"	59.64	421	1.0000(47)	1.0117(9)	1.0013(14)
PU-...-002 7	"	77.09	340	1.0000(47)	1.0128(10)	1.0043(10)
PU-...-003 1	13.0	33.32	787	1.0000(35)	1.0077(9)	1.0077(9)
PU-...-003 4	"	38.12	682	1.0000(35)	1.0106(9)	1.0029(13)
PU-...-003 6	"	44.09	563	1.0000(35)	1.0126(9)	1.0053(12)
Average K_{eff} value over all the variants					1.0112(19)	1.0031(25)

2.3 PLUTONIUM NITRATE SOLUTIONS WITH VARIOUS ^{240}PU CONCENTRATIONS (CROSS SECTION EVALUATION WORKING GROUP. THERMAL REACTOR BENCHMARK COMPILATION. BNL-19302. 1974).

These benchmarks consist of spherical or cylindrical vessels with or without a water reflector.

TABLE 2.3. Critical assemblies containing a plutonium nitrate solution

Assembly	Pu, g/l	H/ ^{239}Pu	^{240}Pu, weight %	Geometry	K_{eff} MCU
PNL- 7		980	0.54	Sphere with refl.	0.9991(20)
PNL- 6		125	4.57	Sphere without refl.	0.9975(19)
PNL-10	28.5	210	8.4	Cylinder without refl.	0.9963(15)
PNL- 9	11.6	910	13.9	Cylinder without refl.	0.9932(11)
PNL-11	40.6	623	42.9	Cylinder with refl.	0.9993(17)
Average K_{eff} value over all the variants					0.9971±0.0025

2.4 MIX OF PLUTONIUM NITRATE AND URANIUM NITRATE SOLUTIONS (NEA/NCS/DOC (95) 03/I. MIX-SOL-THERM-002)

This set of experiments is for a large cylindrical tank that is surrounded with water. The critical height of the "mixed" solutions is measured. The content of ^{240}Pu is equal to 8 %.

TABLE 2.4. Criticality for a mixture of plutonium and uranium nitrate solutions

Experiment number	Pu, g/l	U, g/l	K_{eff}		
			experiment	MCNP	MCU
058	11.88	11.05	1.0000±0.0024	1.0082(2)	1.0005(10)
059	11.73	10.78	1.0000±0.0024	1.0074(5)	1.0012(11)
061	12.19	41.04	1.0000±0.0024	1.0079(4)	1.0013(12)
Average K_{eff} value over all the variants				1.0078±0.0004	1.0010±0.0004

2.5 CONCLUSION

The results of the calculations performed testify to the reliability of MCU-RFFI/A using the DLC/MCUDAT-1.0 library for the analysis of aqueous solutions of plutonium salts. The difference between the results obtained using the MCNP code with the ENDF/B-V library and the MCU results is about 0.7 % in K_{eff}. The MCU-RFFI/A results are generally closer to the experimental value of K_{eff} =1.

3. Validation of the MCU-RFFI/A Code by Comparisons to Experiments on MOX-Water Lattices

Validation of the MCU-RFFI/A code for lattices of uranium-plutonium fuel rods in water was performed as described in reference [4]. The maintenance of high reliability over a wide range of neutron spectra for systems containing plutonium has been demonstrated.

The question of plutonium use in VVER reactors is contingent upon satisfying certain technical criteria. At the present time, experimental data to be used for approval of the design codes for calculations of neutronic parameters of plutonium fueled cores are insufficient. Therefore, the accuracy of the design codes must be assessed on the basis of comparisons to more precise codes based on the Monte Carlo method using evaluated neutron data libraries.

Of course, precise codes also require verification by comparisons to experimental results. Therefore, computational results for the effective multiplication factor, K_{eff}, of heterogeneous critical assemblies with uranium-plutonium fuel and water as moderator and reflector obtained using MCU-RFFI/A are presented later in this section..

Out of the three benchmark sets available in the open literature, two are considered in this paper. These assemblies represent a square lattice of rods placed in a large tank with water. The variables among the assemblies are the fuel composition and water-fuel ratio.

In all of the tables presenting the computational results, the values in parentheses are the statistical variations i.e. one standard deviation. In the case of the average K_{eff} value, the value after K_{eff} is the mean square deviation from the average K_{eff} .

3.1 FUEL WITH a PLUTONIUM CONCENTRATION of 2% WEIGHT.

The assemblies PNL-30 to PNL-35 are described in the Thermal Reactor Benchmark Compilation, BNL-19302 (1974). The fuel rods are made of a mixture of natural uranium dioxide and plutonium dioxide. The ^{240}Pu content in the plutonium is equal to 8%. The diameter of the fuel rods is 0.64 cm and the cladding is zircaloy. Configurations with three different lattice pitches are considered. Pure water and water with boron are used for each configuration. Table 3.1 provides the experimental and computational results.

TABLE 3.1. Assembly with a plutonium concentration of 2 %

Variant	Pitch	V_M/V_F	H/ ^{239}Pu	Boron	K_{eff}	
	cm				Experiment	MCU
30	1.778	2.13	357	-	1.00017	1.000(2)
31	"	"	"	+	1.00006	1.000(2)
32	2.209	7.15	1200	-	1.00028	1.001(2)
33	"	"	"	+	1.00023	1.005(2)
34	2.514	8.13	1365	-	1.00077	1.002(1)
35	"	"	"	+	1.00013	1.008(2)
Average K_{eff} value over all the variants						1.0027±0.0022

3.2 FUEL WITH 20 WEIGHT PERCENT PLUTONIUM

Fast reactor fuel rods are used in these critical assemblies. A full description of the assemblies is available in [3]. The fuel rods are made of a mixture of natural uranium and plutonium (11.5% ^{240}Pu) dioxides. The diameter of the fuel rods is 0.58 cm and stainless steel is the cladding material. Configurations with four different lattice pitches are considered. Table 3.2 provides the experimental and computational results.

TABLE 3.2. Assembly with the plutonium concentration 20 %

Variant	Pitch, cm	V_M/V_F	$H/^{239}Pu$	K_{eff}	
				Experiment	MCU
1	0.9525	1.52	24	1.0000±0.0025	1.0035(10)
2	1.258	3.69	59	1.0000±0.0026	0.9985(10)
3	1.5342	6.18	980	1.0000±0.0032	0.9982(11)
4	1.9050	10.28	163	1.0000±0.0039	0.9997(11)
Average K_{eff} value over all the variants					1.0003±0.0022

3.3 CONCLUSION

The results obtained confirm, that the MCU-RFFI/A code can be recommended for calculation of uranium-plutonium fuel rod lattices in water for a wide range of uranium-plutonium fuel compositions and water-fuel ratios. MCU-RFFI/A can be used for validation of design codes for VVER reactor calculations with plutonium fuel.

4. Comparison to Computational Results for the Benchmark Problems on Burnup of VVER-1000 Lattices and Fuel Assemblies Containing Weapons-Grade Plutonium

Several computational benchmark problems have been proposed within the framework of the Joint Russian-American Program on MOX Fuel Usage in Water Reactors [5]. These problems deal with fuel burnup in VVER-1000 lattices and assemblies containing weapons-grade Pu. A full description of these problems and computational results is published in [6].

The main goal was to verify and validate design codes by intercomparisons of their results, as well as, by comparison to results from more precise calculations . Table 1 provides the K_{eff} and K_0 results obtained using the TVS-M, MCU-RFFI/A (RRC "KI") and WIMS-ABBN (IPPE) codes. The computational results in Table 1 correspond to zero burn up.

The K_0 parameter is defined for a buckled eigenvalue calculation as follows

$$K_0 = \frac{\nu\Sigma^f}{\Sigma^a} \qquad (2)$$

Table 4.1 provides the relative deviations of the results obtained with TVS-M and WIMS-ABBN relative to MCU-RFFI/A. This deviation D is defined as follows

$$D = \frac{K - K_{MCU}}{K_{MCU}} * 100\,\% \qquad (3)$$

TABLE 4.1. Comparison of K_{eff} and K_0 obtained from MCU-RFFI/A, WIMS-ABBN and TVS-M

Var.	State	K_{eff}			K_0		
		MCU	TVS-M	WIMS-ABBN	MCU	TVS-M	WIMS-ABBN
1		S1	1.0639	-0.26	-.63	1.269	-0.27	-.060
2		S3	1.1531	-0.04		1.376	-0.01	
3	V1	S4	1.1033	-0.11	-0.64	1.315	-0.01	-0.45
4		S5	1.1161	-0.09	-0.47	1.332	-0.07	-0.36
5		S6	1.2187	0.13	-0.22	1.370	0.15	-0.21
6		S1	1.0200	0.28	-0.50	1.210	0.36	-0.80
7		S3	1.0614	0.53		1.258	0.72	
8	V2	S4	1.0368	0.39	-0.56	1.229	0.55	-0.76
9		S5	1.0498	0.61	-0.14	1.248	0.54	-0.56
10		S6	1.1786	0.27	-0.06	1.321	0.37	-0.30
11	V3	S6	1.0629	-0.14	-0.09	1.198	-0.36	-0.27
12	V4	S6	1.0004	-0.19	0.03	1.128	-0.51	-0.35
13	V7	S6	1.2501	0.11	0.33	1.403	0.15	0.04
14	V8	S6	1.0828	0.65	-0.45	1.216	0.61	-0.40
15	V9	S6	1.4177	1.17	4.02	1.620	-0.65	1.86
16		S4	1.1172	0.03	-0.78	1.329	0.19	-0.80
17	V11	S2	0.8662	0.04		1.028	0.17	
18		S6	1.2182	0.15	-0.22	1.365	0.21	-0.23
19		S4	1.0600	0.66	-0.46	1.256	0.75	-0.95
20	V12	S2	0.8628	0.81		1.023	0.68	
21		S6	1.1904	0.44	0.20	1.331	0.51	-0.13
22	V13	S6	1.212	0.15		1.358	0.26	

Comments to TABLE 4.1

S1 work state, poisoned	V1 - UO_2 cell
S2 work state with control rods inserted	V2 - MOX cell (W-Pu)
	V3 - spent UO_2 without Fission Products
S3 work state, poisoned	
S4 work state, unpoisoned	V4 - spent UO_2 with Fission Products
S5 hot state	
S6 cold state	V7-V9 - MOX cell (only one Pu isotope)
	V11 - UO_2 fuel assembly
	V12 -Mox fuel assembly
	V13 - multi assembly structure

5. Acknowledgment

The work presented in Sections 2 and 3 was supported by the Russian Foundation for Basic Research through grants N:96-1596797 and N:97-16891.

References

1. M.A. Kalugin and L.V. Maiorov (1996). Application of the Monte Carlo Method for Analyzing the IGR Reactor Experiments. Proceedings of the Topical Meeting "Radiation Protection & Shielding", No. Falmouth, Massachusetts, USA, April 21 -25.

2. N.I. Alexeyev, V.I. Bryzgalov, E.A. Gomin, M.S. Yudkevich (1997). Comparision of the MCU-RFFI/A Calculations with Experimental Data and Other Code Results Applying to the Criticality Safety. Proceedings of the 10th International Topical Meeting on Nuclear Reactor Physics. Moscow, September 2-6.

3   NEA Nuclear Science Committee "International Handbook of Evaluated Criticality Safety Benchmark Experiment". NEA/NCS/DOC (95) 03. Paris 1995.

4. N.I. Alexeyev, V.I. Bryzgalov, E.A. Gomin, M.S. Yudkevich (1997). Validation of the MCU-RFFI/A Code through the Experiments with MOX Water Lattices. Proceedings of the 10th International Topical Meeting on Nuclear Reactor Physics. Moscow, September 2-6.

5. M.A. Kalugin, A.P. Lazarenko, L.V. Maiorov, A.M. Pavlovitchev, V.D. Sidorenko,V.M. Dekusar, A.G.Kalashnikov, E.V. Rojihin, A.M. Tsibulia,M. D.

DeHart, I. Remec, J. C. Gehin, and R. T. Primm (1997). Comparison of Calculational Results for the Benchmark Problems on Burnup of VVER-1000 Lattices and Fuel Assemblies Containing Military-Grade Plutonium. Proceedings of the 10th International Topical Meeting on Nuclear Reactor Physics. Moscow, September 2-6.

6. M.Kalugin, A.P. Lazarenko, L.V. Maiorov, SidorenkoV.M., Dekusar, A.G. Kalashnikov (1997). Calculational Results for Benchmark Problems on Burnup of Lattices and Fuel Assemblies with MOX and U-Gd Fuel. Proceedings of the 7th symposium of AER, Germany, September 23-29 1997, in Hornitz near Zittau

DEVELOPMENT OF NEUTRONIC MODELS FOR TWO TYPES OF REACTORS BASED ON THE SAPFIR PACKAGE OF UNIVERSAL ALGORITHMS

V. ARTEMOV
A. ELSHIN
A. IVANOV
A. KARPOV
V. OBUKHOV,
Yu. SHEMAEV
Science and Research Institute of Technology (NITI)
Sosnovy Bor
Leningrad Region
Russian Federation

V. TEBIN
RRC "Kurchatov Institute"
Moscow
Russian Federation

Abstract

Some experience related to the preparation and validation of neutronic models based on the SAPFIR code package is presented. Development and verification of neutronic parameters for two types of reactors with thermal and intermediate neutron spectra are described.

1. Introduction

The SAPFIR software package (Set of Algorithms and Codes for Investigations of Physical Processes in Reactors) is a system of interconnected segments, modules and libraries for numerical modeling of the neutronic characteristics of reactors (and reactor subregions) based on evaluated nuclear data and multidimensional calculational routines.

T. A. Parish et al. (eds.),
Safety Issues Associated with Plutonium Involvement in the Nuclear Fuel Cycle, 159–172.
© 1999 *Kluwer Academic Publishers. Printed in the Netherlands.*

The SAPFIR [1] package was designed to provide a highly accurate code intended for both benchmarking and research calculation. SAPFIR uses calculational modules and libraries of cross sections developed by experts from several organizations.The SAPFIR package is a general purpose set of software for calculation of the neutronic performance of reactors. The versatility of this package results from the following attributes:

- Only evaluated nuclear data files are utilized as sources of nuclear and physical data and the codes for producing working libraries from the data files are included in the package architecture;
- There are no fundamental restrictions on the composition and geometry of the systems considered; and
- There are no models with adjustable parameters in the important calculational components of the code.

SAPFIR has achieved its goals for accuracy and versatility through implementation of the following techniques/features:

- Adequate representation of basic nuclear cross section data in the working libraries [2];
- Generalized subgroup approach (GSA) in the resonance energy range [3];
- Thermalization modeling using 40-microgroups [4];
- Burnup considering nearly 70 isotopes [5]; and
- Calculations of fast collision probabilities (FCP) by the method given in [6] based on the geometrical module SCG-5 [7], [8].

In addition, the SAPFIR package includes a set of segments for benchmark calculations using both the Russian database of the "Kurchatov Institute" [9] and ENDF/B type nuclear data libraries. This makes it possible to prepare a set of test tasks during verification. These tasks are based on the following:

- Solution of the neutron moderation equation with detailed representation of the cross sections in the resolved resonance energy range;
- Solution of the neutron transport equation in cells (and reactors) by both Monte Carlo and discrete ordinates methods;
- Solution of neutron multidimensional diffusion problems using many energy groups and fine finite difference meshes.

2. Model Development

To be applicable to computational problems involving a specific type of nuclear reactor, a code system is assembled from modules and segments. The code system is then verified and submitted to the Russian Nuclear Inspectorate (GAN) for registration and certification. This paper describes the licensing procedure for code systems based on the experience with SAPFIR, from the preparation of a working library, all the way to approval by GAN. As an example, code systems for two types of reactors, VVERs and nuclear propulsion power plants, are reviewed.

The first stage in licensing a code system is the preparation of multigroup libraries of neutron cross sections based on detailed nuclear data. Working libraries are prepared by a separate code unit of SAPFIR, which in combination with the well known NJOY code produces libraries in TEMBR and TEPKON format for the epithermal and thermal neutron energy ranges, respectively. Parameters for GSA are calculated for resonance isotopes with 12 to 23 groups (BNAB). The working library for nuclear propulsion power plant analyses, as compared to the working library for VVER analyses, is extended to include data for additional isotopes that are particular to such reactors.

At the same time as they are generated, the constants obtained are preliminarily tested. In the resolved resonance energy range, the results obtained by solving the moderation equation with a detailed point by point representation of the neutron cross sections, are used for this purpose. The final verification of the libraries is carried out through comparisons to numerical and experimental benchmarks.

The second stage of the code system assembly process involves the formulation of codes/data, based on the SAPFIR package, to perform a "cell" analysis. The main program is prepared based on the type of reactor. This program calls segments in a specific sequence. The formation and installation task is performed using the SAPFIR package's systems support. With this support, the tasks for a required configuration (segments set, dimensions) is formed, and data input/output and exchange among the segments is built. For example, the core analysis code system for nuclear propulsion reactors, is composed of a classical element and contains two key units, i.e., the cell calculation code SAPFIR_VVR95, with the library, BNAB-78/C-95, and the universal three-dimensional diffusion mesh code RC10.

Some SAPFIR package segments not needed for VVER type calculations are included in the nuclear propulsion reactor code system. In particular, the FCP calculation employs a universal geometric module for simulating cell geometry of any complexity. In order to model burnable poison, different modules are utilised which simulate the "surface burnup" of an absorber rod. The code system for nuclear propulsion reactors also uses specialised segments for thermal-hydraulic calculations.

Appreciably different modules are used in the code systems which model the elements that simulate the control and protection systems, ie., clusters and "traps" for VVER-1000s and VVER-440s, respectively, and absorber rods located in the interchannel space for nuclear propulsion reactors. For VVERs, one also needs to model the replacement of assemblies during refuelling and the use of soluble absorber.

3. Verification and Validation (V&V)

The final, and highly important, stage in a code system's development is verification and validation. This is discussed in the following sections. Codes systems need to be verified with regard to the recommendations made by GAN. A set of tests was selected based on earlier verification activities such as the one described in [10]

- Checking the nuclear data's validity;
- Checking that mathematical models and calculational algorithms are adequate for the physical problem to which they are applied;
- Definition of the application area for the models and algorithms implemented;
- Assessment of the accuracy of calculated neutron physics characteristics important to reactor safety.

For practical applications, the last task in the list, is very important. However, in order to accomplish it, the first three tasks must be accomplished first. Indeed, the final error in the computed integral characteristics of a reactor includes a number of components such as:

- Uncertainties in the nuclear data utilized;
- Inaccuracies of models for description of physical conditions assumed used in the execution of the spectral task (this task involves modeling neutron transport during moderation, resonance absorption, and thermalization in a reactor cell's geometry);
- Inaccuracies due to homogenization and neutron diffusion models of the reactor core;
- Inaccuracies of models for description of the feedback related to temperature and the change in isotope composition due to reactor burnup and poisoning; and
- Uncertainties in initial data due to technological limitations.

In view of the previous considerations, the experimental data employed for code validation were divided into groups according to their type. In addition, parallel numerical tests were carried out at all of the V&V stages. For these tests, results ("Calculational Benchmarks") were compared to those from reference codes of the MCNP or MCU type, as well as, the reference algorithms of the SAPFIR package.

As a result, the V&V of the SAPFIR_VVR95-RC code system were performed based on a system of tests as follows:

1. Calculational and experimental test problems of the "Nuclear Data Benchmark" type intended for justification of the nuclear data and models utilised in the "spectral" tasks;
2. Calculational and experimental test problems of the "Separate Effects Benchmark" type intended for justification of specific neutron physics models taking into account changes in isotope composition, fuel and coolant temperature, absorber presence, etc.;
3. Calculational and experimental test problems based on fuel elements and assemblies with properties similar to certain reactor types ("Core Physics Benchmarks"). Experiments and test problems from this group should be

selected such that maximum possible coverage of parametric variations in the fuel rods, assemblies, control absorbers and burnable poisons is ensured;

4. Tests problems to compare to measurements of neutron physics characteristics on full scale cores at the beginning of an operating cycle (startup). This group includes experiments for different reactor types with changes in core state (fuel addition, heat up, removal and insertion of control rod groups, etc.). For V&V, comparisons with the results from direct measurements were mostly utilised, such as, critical values of boron concentration, measured differential and integral worths of control rods, temperature coefficients of reactivity, and power distributions in individual fuel assemblies;

5. Comparisons to measured neutron physics characteristics on full scale cores of operating reactors taking into account full power operation, neutron poison positions and burnup. This group includes experiments based on reactor operation at full power and at different power levels, as well as, measurements at "zero" power at different times during a cycle.

3.1 NUMERICAL TESTS OF THE "NUCLEAR DATA BENCHMARK" TYPE

Some V&V results for several typical tests are given below. Table 1 shows the results of calculations performed with SAPFIR_VVR95-RC based on the BNAB-78/C-95 library in comparison to calculational results obtained based on the ENDF/B-V library [11].

3.2 NUMERICAL AND EXPERIMENTAL TESTS OF THE "SEPARATE EFFECTS BENCHMARK" TYPE

As an example of the test problems in this group, calculation of the Doppler effect, void effect and burnup effects are presented.

3.2.1. *Doppler effect modeling using SAPFIR_VVR95-RC*

This section presents results of Doppler effect calculations for the mathematical test problem described in [12,13]. Table 2 shows the results of a k_{inf} calculation at different temperatures. Table 3 presents the reactivity change for the given temperature range.

164

TABLE 1. Spectral indexes in the numerical tests

Test	Code	k_∞	$\rho 28$	$\delta 25$	$\delta 28$	CR
NB-1	Monte Carlo	1.1449(.14)	1.363(.6)	0.0803(1.1)	0.0722(.6)	0.798(.4)
	SAPFIR_ VVR95-RC	1.1461	1.3573	0.0795	0.0689	0.7949
NB-2 a	Monte Carlo	1.1748	2.612	1.51-1	2.97-1	2.148
	SAPFIR_ VVR95-RC	1.1755	2.559	1.48-1	2.81-1	2.115
NB-4	Monte Carlo	1.3424(.26)	2.654(.6)	0.159(.6)	0.0617(.8)	0.549(.3)
	SAPFIR_ VVR95-RC	1.3433	2.647	0.1547	0.0599	0.5496
NB-5	Monte Carlo	1.1456(.15)	8.503(.8)	0.5480(1.7)	0.133(.5)	1.006(.3)
	SAPFIR_ VVR95-RC	1.1395	8.490	0.5443	0.1295	1.0065

a The uncertainty of these Monte Carlo values was not available

TABLE 2. Multiplication factor (k_{inf}) versus temperature

Enrichment (wt%)	Temperature, K		MCNP	SAPFIR _VVR95- RC	Deviation in reactivity
	Water	Fuel			
0.711	300	300	---	0.6077	
1.6	300	300	0.9177(33)	0.9184	0.08%
2.4	300	300	1.0756(37)	1.0733	-0.20%
3.1	300	300	1.1597(42)	1.1636	0.29%
0.711	600	600	0.6638(6)	0.6653	0.34%
1.6	600	600	0.9581(6)	0.9607	0.28%
2.4	600	600	1.0961(7)	1.0987	0.22%
3.1	600	600	1.1747(6)	1.1768	0.15%
3.9	600	600	1.2379(6)	1.2400	0.14%
0.711	600	900	0.6567	0.6575	0.19%
1.6	600	900	0.9484	0.9507	0.26%
2.4	600	900	1.0864	1.0881	0.14%
3.1	600	900	1.1641	1.1658	0.13%
3.9	600	900	1.2271	1.2284	0.09%

TABLE 3   Doppler effect

Enrichment (wt%)	Temperature range	MCNP	SAPFIR _VVR95-RC
0.711	600-900	-0.0163(23)	-0.0178
1.6	600-900	-0.0108(9)	-0.0110
2.4	600-900	-0.0081(8)	-0.0089
3.1	600-900	-0.0078(7)	-0.0080
3.9	600-900	-0.0071(6)	-0.0076

3.2.2. The NEACRP HCLWR burnup benchmark problem

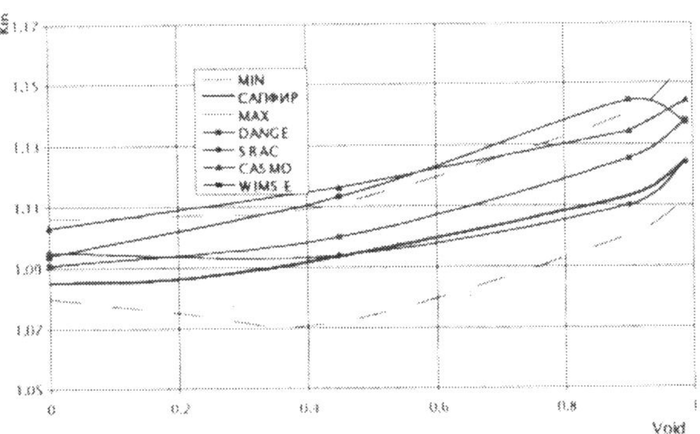

Fig. 1. Multiplication factor vs void fraction in cell with V_m/V_f=0.6.

The test calculation of burnup with uranium - plutonium fuel [14] in tight lattices is a stringent test for spectral codes. This test is of special interest because calculated results from many well-known neutron physics codes are available.

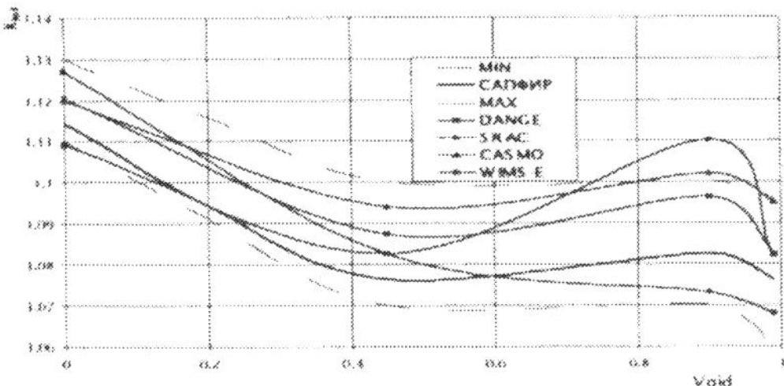

Fig. 2. Multiplication factor vs void fraction in cell with V_m/V_f=1.1

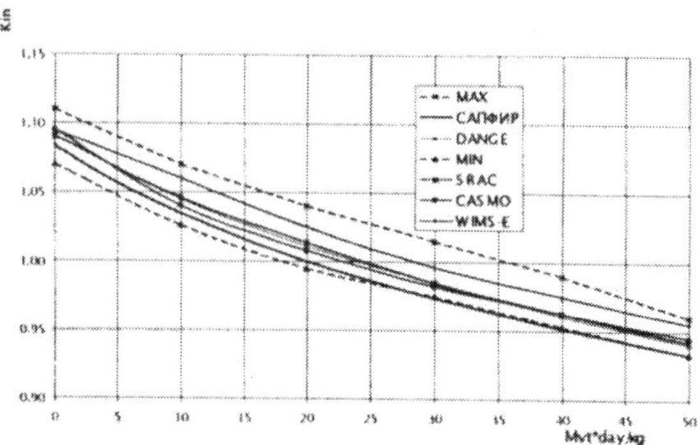

Fig. 3. Multiplication factor vs burnup in cell with V_m/V_f=0.6.

The SAPFIR_VVR95-RC calculations were carried out for different water densities (void fraction) in the cells at the beginning of cycle. For the initial void fraction, the burnups are calculated up to 50 GWD/MT. Figures 1-4 show the calculated results from SAPFIR_VVR95-RC in comparison to the results from other codes. Table 4 compares the concentrations of the key isotopes as obtained from burnup calculations using SAPFIR_VVR95-RC and DANGE (70 groups, JEF-1 library, block factors in resonance region).

3.2.3.   *Power reactor depletion experiments*

Calculated results for burnup dependent Pu isotopic ratios (^{239}Pu/^{240}Pu, ^{240}Pu/^{241}Pu, ^{241}Pu /^{242}Pu) are shown in Fig. 5 in comparison to experimental data [15] from the Yankee reactor.

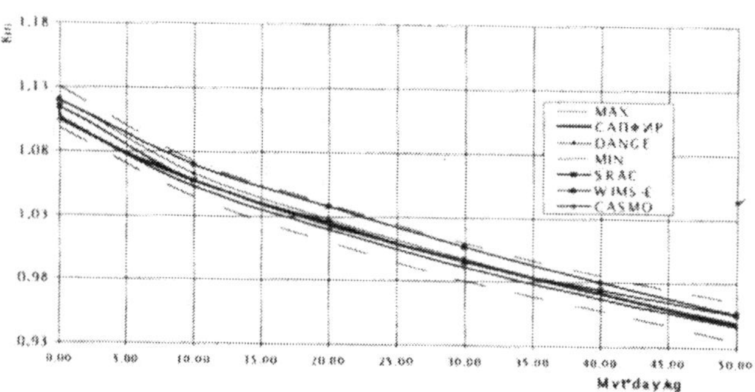

Fig. 4. Multiplication factor vs void fraction in cell with V_m/V_f=1.1.

TABLE 4 Isotopic concentrations

Cell 1 (Vm/Vf=0.6)						
Burnup	30 GWt*day/t			50 GWt*day/t		
Isotope	SAPFIR	DANGE	errot(%)	SAPFIR	DANGE	error(%)
235U	4.096E-05	4.078E-05	0.44%	3.086E-05	3.086E-05	0.00%
238U	1.959E-02	1.961E-02	-0.12%	1.914E-02	1.917E-02	-0.19%
236U	5.432E-06	5.602E-06	-3.04%	7.435E-06	7.642E-06	-2.71%
239Pu	1.446E-03	1.437E-03	0.63%	1.387E-03	1.376E-03	0.78%
240Pu	6.812E-04	6.877E-04	-0.94%	6.740E-04	6.874E-04	-1.95%
241Pu	2.798E-04	2.823E-04	-0.89%	2.792E-04	2.803E-04	-0.40%
242Pu	2.017E-04	1.974E-04	2.16%	1.996E-04	1.896E-04	5.28%
Cell 2 (Vm/Vf=1.1)						
235U	4.236E-05	4.240E-05	-0.09%	3.234E-05	3.239E-05	-0.15%
238U	1.996E-02	2.005E-02	-0.44%	1.958E-02	1.968E-02	-0.48%
236U	5.130E-06	5.233E-06	-1.98%	7.127E-06	7.261E-06	-1.84%
239Pu	1.124E-03	1.124E-03	0.02%	1.003E-03	1.007E-03	-0.38%
240Pu	5.607E-04	5.675E-04	-1.20%	5.247E-04	5.389E-04	-2.65%
241Pu	2.921E-04	2.897E-04	0.81%	3.022E-04	2.976E-04	1.53%
242Pu	1.811E-04	1.750E-04	3.44%	1.858E-04	1.726E-04	7.68%

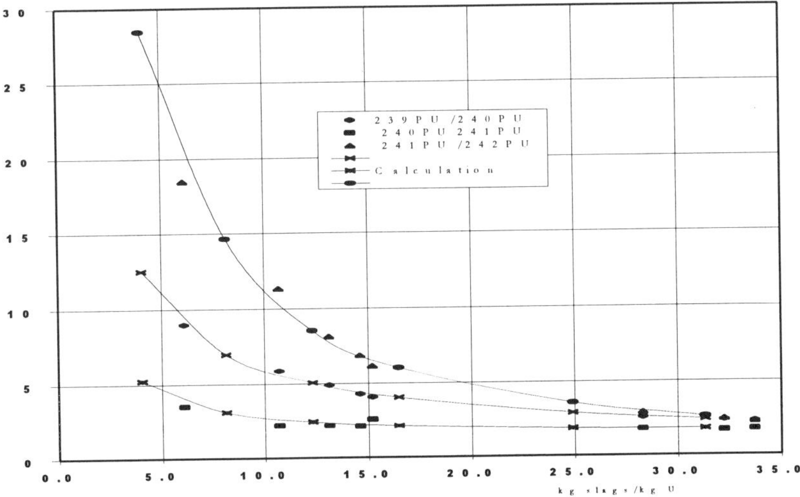

Fig. 5a Plutonium isotopic ratios vs burnup

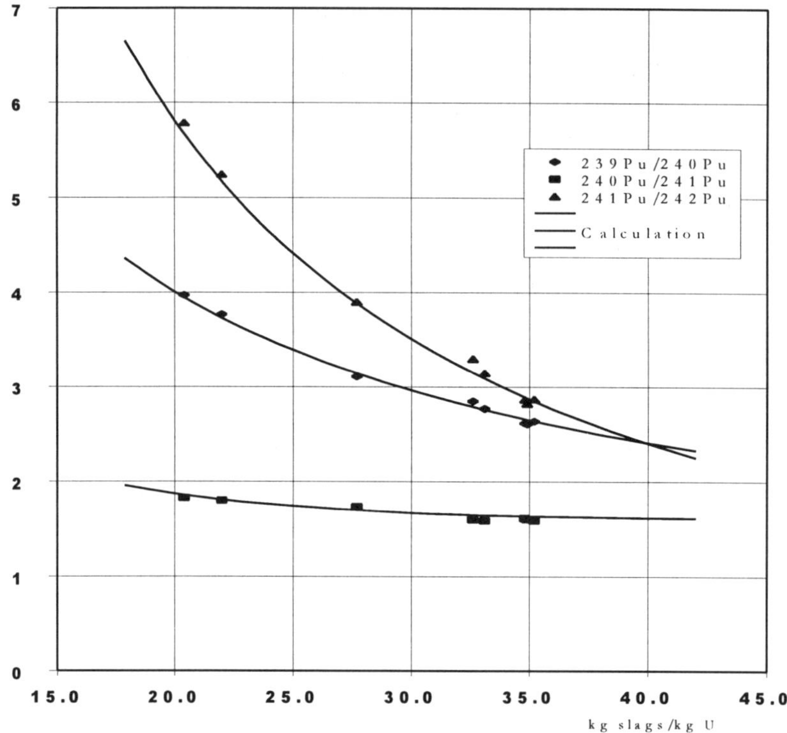

Fig. 5b Plutonium isotopic ratios vs burnup

TABLE 2. Light water reactor features considered in the development and V&V of the SAPFIR_VVR95-RC code system

Experiment and facility designs	Test number (arbitrary)							
	1	2	3	4	5	6	7	8
1. Fuel element:								
1.1. Rods (pins)	+	+	+	+		+	+	+
1.2. Ring					+	+		
2. Fuel:								
2.1. Metal				+		+		
2.2. Ceramic.	+				+	+		
2.3. Disperse.		+	+			+	+	+
3. Assembly:								
3.1. Channel-type		+	+	+	+		+	+
3.2. Hexagon (Cassete)	+							
3.3. Assembly with different fuel elements						+		
4. Absorber elements :								
4.1. Control rods							+	
4.2. Control rods with plates						+		
4.3. Cluster	+							
4.4. Single rod								+
5. Absorbers:								
5.1. Pins (rods)						+	+	+
5.2. Ring						+		
6. Burnable poisons:								
6.1. Rods			+				+	+
6.2. Bars			+				+	+
6.3. Assembly elements made of structural materials with poison		+						
6.4. Burnable poison mixed with fuel	+							
7. Lattice pitch variation			+					
8. Heating to the hot state		+			+	+	+	+
9. Drastic change in the coolant density (up to complete dry out) in fuel assembly				+				
Deviations from the benchmark	1	2	3	4	5	6	7	8
1. Multiplication factor								
1.1. Mean square deviation (%)	0.1	0.1	0.1	0.2	0.2	0.2	0.4	0.4
1.2. Maximum deviation (%)	0.2	0.3	0.3	0.6	0.3	0.4	0.7	0.7
2. Power density								
2.1. Mean square deviation (%)	1.7	1.8	2.3	1.4	0.2	1.2	1.4	1.6
2.2. Maximum deviation(%)	4.0	4.0	5.0	3.0	0.3	2.6	2.7	2.8

3.3.   FINAL STEPS OF V&V

V&V matrices for the reactor types under consideration partially intersect and partially supplement each other. In some cases, the most representative benchmark tasks are VVER specific tests such as:

- V&V of models for calculating the neutron spectrum and resonance absorption, most of all for ^{238}U as a function, lattice pitch and fuel temperature (Doppler effect);
- V&V of models for fuel burnup and plutonium build-up in cell calculations; and
- V&V of algorithms for cell and reactor calculational codes simulating refuelling with reactivity compensation by a soluble absorber.

In other cases, the most stringent tests of the computer code system are those imposed by tasks for the nuclear propulsion reactor:

- V&V of burnup models for "heavy" absorbers;
- V&V of algorithms for solving the neutron transport equation for cells with complicated geometry and non-uniformities, such as, heavy absorbers, different types of fuel rods, etc.;
- V&V of models and algorithms for calculation of the power density in cores that are non-uniform radially and axially;
- Assessment of the validity of calculated reactivity effects; and
V&V of joint neutron physics and thermal-hydraulic models.

Thus, it can be seen that an integrated verification matrix allows for testing of individual algorithms and complete code systems at a qualitatively new level. In practice, all of the VVER specific algorithms remained unchanged within the overall structure of the SAPFIR system when developing/applying SAPFIR_VVR95-RC for propulsion reactors. This version was the last to be developed. The verification was conducted employing the integrated verification matrix. The characteristics of the code system are summarized in Table 6.

TABLE 3. Experiments with full-scale cores used in V&V of SAPFIR_VVR95-RC

Measured parameters	Without burnup	With burnup
1. Core criticality in experiments with fuel unloading	1.0003 ±0.0008	
2. Core criticality in the cold state	1.0009 ±0.0008	1.0001 ±0.0008
3. Core criticality in the hot state	1.0001 ±0.0007	0.9999 ±0.0016
4. Core criticality in experiments with core heat up to operational conditions	0.9986 ±0.0007	1.0002 ±0.0015
5. Differential worth of control rods under operation condition	5%	5%
6. Differential worth of control rods during core heat up	7%	7%
7. Temperature reactivity factor	10%	10%
8. Core criticality in experiments with recompensation of control rods	1.0006 ±0.0010	0.9999 ±0.0015
9. Differential worth of control rods at recompensation of control rods	7%	10%
10. Core criticality in experiments with recompensation of some of the control rods	1.0017 ±0.0022	
11. Differential worth of some of the control rods at recompensation	10%	
12. Worth of control rods in the experiments with drop of some of the control rods (core subcriticality).	5%	5%
13. Worth of scram rods in the experiments with scram (rod drop)	10%	
14. Core criticality in the experiments with different reactor power operation modes		0.9988 ±0.0022
15. Core criticality in the experiments with reactor poisoning and overpoisoning with xenon	1.0006 ±0.0015	0.9990 ±0.0006
16. Differential worth of control rods in the experiments with reactor poisoning and overpoisoning with xenon	7%	
17. Power density peaking factor (k_r, k_z)	3%	
18. Core power density	6%	
19. Axial power density	5%	7%
20. Fuel elements power density	7%	

4. Conclusions

Software for calculations on other types of reactors can be licensed using a similar scheme. Few additions to the set of functional codes in SAPFIR are expected to be required. For example, a code system for RBMK calculations has been prepared and it did not require changes to the nuclear data working library or to the basic codes. Hence, the software only had to be verified.

References

1. Tabin I, Buchov V, Sergeev V. Ivanov A et al (1985) Software package SAPFIR for reactor cells calculcation:: VANT, series Nuclear Reactors Physics and Engineering, 4, pp. 68-71.
2. Karpov, A, Tabin, V. (1995) Many group neutron data preparation system for SAPFIR package with NJOY-B30 code system. Proceedings of Neutronic-95 seminar, Obninsk, Russia.
3. Tabin V, Judkevich I, (1985) generalized subgroup approach to resonance absorption calculation: Atomic energy, v. 59, iss. 2, p. 96.
4. Gomin E., Maiorov L. (1982) Code TERMAC for group neutron cross-sections in thermalization region VANT, series Nuclear Reactor Physics and Engineering, 5 (27), n. 70.
5. Pisarev P. (1992) Modules OTRAWA and SVP for burnup calculations in BMVC and SAPFIR codes, Third conference proceeding "Nuclear Technology Tomorrow", Leningrad.
6. Gomin E., Maiorov L. (1982) About FCP calculation in complex geometry systems, VANT, series Nuclear Reactors Physics and Engineering 8 (21), p. 62-69.
7. Alexeev N., Gurevich I. (1993) Geometry module SCG-5. Preprint IAE-5616/4.
8. Ivanov A. (1984) Bank of neutron constants of Institute of Kurchatov, Institute of Atomic Energy: VANT, series Nuclear Data 6 (43), p. 55-56.
9. Judkevich I. (1984) Bank of neutron constants of Institute of Kurchatov Institute of Atomic Energy: VANT, series Nuclear data, 5 (59), p. 3.
10. ArtemovV., Ivanov A, Piscarev A, Obuchov V., Shemaev Yu. (1996) The set of tests and V&V experience calculation core code system SAPFIR_VVR95-RC on water-water propulsion reactors facilities. Proceedings of Neutronic-96 seminar, Obninsk, Russia.
11. M.L. Williams, R.Q.Wright et al. (1985) Analysis of Thermal Reactor Benchmarks with Design Codes Based on ENDF/B-V Data. Nuclear Technology, vol. 71, pp. 386-401
12. Mosteller R.D. etc. (1989) Benchmarking CELL-2 Using a Monte Carlo Method, NSAC-136.
13. Mosteller R.D. etc. (1991) Benchmark calculations for the Doppler Coefficient of Reactivity, Nuclear Science and Engineering 8, p. 437.
14. Y.Ishiguro, H.Akie, H.Takano (1988) Preliminary Report of HCLWR Cell Burnup Benchmark Calculation, NEACRP-A-849
15. Stepanov A., Makarova A. et all (1983) Burnup and isotope composition determination VVER-440 spent fuel, Atomic energy, vol. 55, is. 3

PERFORMANCE MODELING FOR WEAPONS MOX FUEL IN LIGHT WATER REACTORS

K. L. PEDDICORD
J. ALVIS Jr.
Department of Nuclear Engineering
Texas A&M University
College Station, Texas 77843-3133
United States

Abstract

The thermal, mechanical and chemical performance of mixed uranium-plutonium oxide fuel manufactured from excess weapons plutonium is of prime importance to the implementation of this disposition option. This paper assesses some of the technical questions associated with fuel performance and the modeling of these phenomena.

1. Introduction

A principal option being considered in both the Russian Federation and the United States for the disposition of excess plutonium from disassembled nuclear warheads is the burning of this material as mixed uranium-plutonium oxide (MOX) fuel in light water reactors (LWR's). The Russian Federation's Ministry for Atomic Energy (Minatom) has made a policy statement that excess weapons plutonium is a national treasure which will be burned to recover its energy value [1]. Conversely, the US Department of Energy has announced its Record of Decision [2] in which MOX and immobilization with radioactive fission products are identified as the two disposition options . In both cases, the objective[3] is to achieve the spent fuel standard as identified by the US National Academy of Sciences [4] study on the disposition of weapons plutonium.

To utilize MOX fuel in light water reactors, a number of technical issues must be addressed with respect to the performance of this fuel under typical power reactor

173

T. A. Parish et al. (eds.),
Safety Issues Associated with Plutonium Involvement in the Nuclear Fuel Cycle, 173–177.
© 1999 *Kluwer Academic Publishers. Printed in the Netherlands.*

conditions. To do this, a variety of fuel performance models are utilized which seek to describe the behavior of the fuel during irradiation.

2. Background

Somewhat ironically, the two countries which are planning to use the MOX option for plutonium disposition currently have only a modest technical basis from which to draw. In an executive decision in 1979, the US decided to forego reprocessing of spent nuclear fuel because of the unfavorable economics of MOX fuel due to abundant global uranium resources, and as a nonproliferation initiative, to avoid separating plutonium which might be used to manufacture weapons. As a result during the subsequent time, the technical infrastructure in the U.S. which had been built up for MOX use in light water reactors has eroded.

In contrast, Russia has chosen to pursue a closed fuel cycle including the reprocessing of LWR fuel. However, the early emphasis of the Russian program had been on the use of MOX fuel in fast reactors and no MOX experience has been accrued as yet in Russia for MOX use in LWR's.

In the US, a number of the 110 operating LWR's are qualified for the weapons plutonium disposition mission. Some 3 to 7 are projected to be needed to meet the requirements. No further reactor construction is anticipated in the United States. In Russia, currently only 4 of the 7 VVER-1000 reactors have been judged appropriate for plutonium disposition. Russia is also considering the use of the BN-600 for this purpose. Future reactors are projected for construction, both LWR's and fast reactors, which could serve this role as well. The net effect of this history is that all of the current operating experience with MOX fuel resides in Western Europe and Japan. However, virtually all irradiated MOX fuel has been manufactured with reactor grade rather than weapons-grade plutonium.

3. Fuel Performance Issues

A number of phenomena play important roles in the performance of LWR nuclear fuels. Virtually all mechanisms of interest are regarded as being thermally driven. As a result, the thermal behavior of LWR fuel receives much attention. To adequately understand the thermal performance, a precise knowledge of the thermal conductivity is needed, especially as a function of fuel burnup. In addition, the behavior of the gas-filled gap separating the fuel pellets and the clad is of major importance.

A direct result of the thermal behavior of the fuel matrix is the release of fission products to the internal volume of the fuel pin. Because of the generally lower thermal conductivity of the fission product gases, these in turn impact the thermal performance. Additional release of fission product gas also leads to increasing internal pin pressure. The fission product gas behavior is of major importance because in the event of even a small "pin hole" failure of the clad, the release of fission product gas to the primary coolant loop represents the principal potential radiological impact of reactor operation under non-accident conditions. In addition to the fission product gas behavior, other solid fission products can play a role in determining the gap thermal conductance because of their corrosive effect on the zircaloy clad. As a result, the clad response, and integrity of the clad in turn, is often governed by the thermal conditions, fission gas product pressure, and the solid fission product behavior.

This situation is further complicated because of the trend towards higher fuel burnups. It is economically attractive to utilities to operate with longer refueling intervals, to reduce refueling outages, and to reach higher fuel burnups. With the corresponding increase in the fission product inventory, the phenomena summarized above play increasingly greater roles. Because of this interplay, data suggest that the release of fission product gas for MOX fuel under high burnup conditions, i.e., above approximately 45,000 MWD/MTM, increases over that for low enriched uranium fuel at the same burnups [5], see Figure 1. The physical mechanisms which produce this result may include differences in the high burnup thermal properties of the two fuel types, or alternatively, may be linked to the manufacturing processes used in producing MOX as opposed to LEU fuel. These issues take on importance because it will be necessary to fit them into the current operational strategies of the utilities owning the reactors. While this is assumed to be of greater importance to the US disposition efforts, these issues could also play a role in Russian LWR's as well.

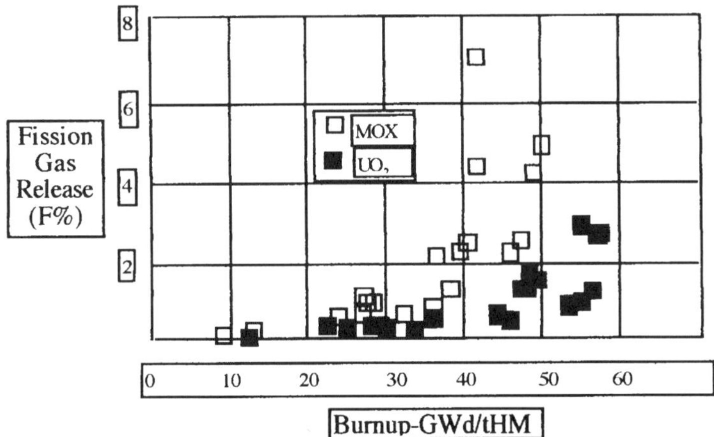

Fig. 1: Fission Gas Release of MOX Fuel Rods Compared to UO_2 Rods as Function of Burnup[5]

176

4. Fuel Performance Modeling

Prediction of LWR fuel performance is done with extensive computer codes which seek to model several mechanisms which operate on vastly different time scales. In addition to being coupled, the thermal, mechanical, chemical and neutronic phenomena are often not fully characterized and cannot all be modeled entirely for the MOX fuel from excess weapons Pu on the basis of first principles of physics. As a result, many approximations are necessary, and the codes must be benchmarked to the maximum extent possible against irradiation data.

A number of codes exist to model the irradiation behavior of oxide fuel, although many fewer of these codes have been developed for MOX fuel. None of the current LWR codes in the US and Russia meet this need. In addition, in Europe where MOX is used, the target burnups for the fuel are typically lower than those used in the US, although this trend is changing. As a result, much attention is now being given to extending the fuel performance models of MOX fuel to higher burnup regimes, and to test these models against available data. The observed higher release of fission product gas from the fuel pellets for MOX fuel must be well understood so that this behavior can be accommodated in the fuel and cladding design. In addition, any potential role in the irradiation behavior from the differing fuel manufacturing methods must also be well understood. The future work in fuel modeling will focus on these issues. Finally, this assumes that the other unique features of weapons plutonium metal such as the presence of gallium or other trace elements will be mitigated so that the source material for fuel manufacture is essentially identical, except for isotopics, to reprocessed reactor grade plutonium.

5. Conclusions

For utilities to be able to irradiate MOX fuel manufactured from weapons plutonium, the irradiation behavior of this fuel must be well understood. In addition, suitable predictive models must be available to assure that MOX fuel from weapons Pu can be irradiated to the target burnups while meeting expected fuel performance and safety standards. Work is needed to draw upon existing data, to add to the knowledge base through appropriate experiments, and to develop additional fuel performance models to meet these requirements.

6. Acknowledgment

This paper was prepared with the support of the US Department of Energy (DOE), Cooperative Agreement No. DE-FC04-95AL85832. However, any opinions, findings, conclusions, or recommendations expressed herein are those of the authors and do not necessarily reflect the views of DOE. This work was conducted through the Amarillo National Resource Center for Plutonium.

References

1. Murogov, Second International Policy Forum: Management and Disposition of Nuclear Weapons Materials, Landsdowne, VA, March 21-24, 1995.
2. DOE Record of Decision on the Storage and Disposition Weapons-Usable Fissile Materials, January 14, 1997.
3. Joint US/Russian Plutonium Disposition Study, September, 1996.
4. "Management and Disposition of Excess Weapons Plutonium," US National Academy of Sciences, October, 1994.
5. Trotabas, P. Blanpain, N. Weakel, D. Haas, and P. Menut, "La R&D en soutien de la conception et du comportement du MOX en reacteur: gaz de fission, interaction pastillesgaine," *SFEN Technical Meeting on the R&D of MOX Fuel for Pressurized Water Reactors,* Paris, France, October 27, 1995.

NUCLEAR POWER SYSTEMS USING FAST REACTORS TO REDUCE LONG-LIVED WASTES

V.I. MATVEEV
I. Y. KRIVITSKI
A.G. TSIKUNOV
Institute of Physics and Power Engineering
Obninsk, Russia

Abstract

In this paper, a concept for a nuclear power system with a closed fuel cycle is considered. The system consists of thermal and fast reactors and achieves utilisation of practically all of the actinides produced by both types of reactors. The important characteristics of fast reactor cores designed for effective actinide burning are presented. Limitations on the fast reactor's fresh fuel composition, based on heat generation and external radiation characteristics, are given. The results of computational studies to determine the fuel cycle characteristics (assuming repeated fuel recycling) of a nuclear power system consisting of VVER-1000 and BN-800 type reactors are presented. Calculations were carried out for different types of BN-800 reactor cores, both with (oxide) fuel of a higher than normal enrichment and with fuel not containing U-238. Burnup levels and decay times for the spent fuel were varied.

1. Introduction

The most pressing problem associated with the use of nuclear power is the accumulation of high level waste (HLW) in the form of fission products and actinides. The latter are dominated by plutonium and a group of so-called minor actinides, like neptunium, americium and curium. Possibilities for addressing the HLW problem that are currently under consideration include 1) deep geologic storage and 2) "incineration" of the actinides and long-lived fission fragments in reactors.

The first solution, despite its apparent simplicity, is not easy. A quite well-

T. A. Parish et al. (eds.),
Safety Issues Associated with Plutonium Involvement in the Nuclear Fuel Cycle, 179–191.
© 1999 *Kluwer Academic Publishers. Printed in the Netherlands.*

180

founded distrust exists when it comes to subjects such as, waste form integrity during storage, stability of rock formations, possibility to provide sealing of inlets, possibility to prevent water intrusion from adjacent rocks, etc., due to the time scales involved. It therefore seems that the second solution may provide a more reliable control method than opposed to the first. This paper considers some aspects of solving the HLW problem by burning actinides in reactors.

Earlier studies have shown that a totally closed fuel cycle can not be accomplished, for practical purposes, with a nuclear power system consisting of only thermal reactors. The increase of the feed fuel enrichments at every recycle, the significant increase of the minor actinide fraction, and the increase of the decay heat production in feed fuel are all beyond the existing technology for (mixed) fuel production. Also, in addition to limiting the fractions of the even-numbered isotopes in the plutonium produced, the separation of minor actinides is necessary. And, hence, the disposal problem for these long-lived wastes is not solved.

The capabilities of fast reactors differ substantially from those of thermal reactors on this issue. Fundamental features of fast reactors, linked to the neutron cross sections of the actinides for high-energy neutrons, allow for effective burning. Theoretical studies have shown that fast reactors can operate with a closed fuel cycle enabling almost complete utilisation of the actinides produced. Even n this case, as a result of efficiencies of chemical separation processes a small amount of minor actinides are not recyclable and are lost to the waste stream during reprocessing. For modern radiochemistry, this non-recyclable part is some 0.5 % of the total quantity of the actinides produced. This means that the quantity of, for example, plutonium, committed to wastes can be decreased hundreds of times, as compared with an open fuel cycle. In the future, as chemical processing methods are improved, the decrease may reach much higher levels.

Thus, solution of the actinide burning problem can be achieved using fast reactors and an appropriate mix of thermal reactors within an overall nuclear power system.

2. Concept for Fast Actinide Burner Reactors

Irrespective of breeding fuel, the introduction of fast reactors into a nuclear power system will permit the carrying out of effective burning of the actinides. In the overall system, the fast reactors function primarily as waste burners while the thermal reactors function primarily as electricity producers. Some of the design characteristics of conventional fast breeder reactors needed to be converted to better conform to the actinide burner reactor role. The conversion requirements have been identified at a conceptual level and include the following [1], [2]:

- replacement of breeding blankets by non-breeding blankets;
- increase in the fuel enrichment; and
- use of fuel with an inert material replacing U-238.

Replacement of radial breeding blankets with non-breeding blankets does not lead to any new problems, and is limited primarily by the technical specifications of a particular reactor. Well known materials, used in fast reactor technology, such as stainless steel and boron carbide (with natural B-10 enrichment), can be considered as candidate materials for non-breeding blankets. Replacement of axial breeding zones, which are included within the fuel elements themselves, involves more technical difficulties than the replacement of radial breeding blankets. Acceptable technical solutions have been found both for the already operational BN-600 reactor and for the BN-800 reactor design.

A more complicated problem appears to be the required increase in the fuel enrichment. Oxide fuel, which is considered the leading candidate fuel type for actinide burning is well mastered from a technological viewpoint, but has a limit on enrichment determined by solubility. The available data suggest that one can reach enrichment levels of 40 to 45 % although this will require additional technological improvements. This enrichment level can provide effective actinide burning characteristics. As an example, a fast reactor with such a fuel enrichment at a power of 1 GW is capable of burning 500 kg of plutonium per year (at a capacity factor (CF) = .80). What are the changes needed if the fuel enrichment used in the reactor is to be increased? The principal features are as follows [1], [3]:

- introduction of neutron absorbers into the core;
- a decrease in the fuel volume fraction; and
- introduction of absorbing blankets.

An engineering optimisation study can find the most appropriate core variants both for operational fast reactors and for fast reactors still under design.

The highest actinide burning efficiency can be obtained using fuel that does not contain U-238. Reactors with such fuel (power of 1 GW and CF=0.8) can reach an upper limit for plutonium burning of about 750 kg/year. Some fuel compositions that do not contain U-238 are under development. These use inert matrix materials, like magnesium oxide, zirconium carbide, aluminium nitride etc., for placing the actinides within the fuel elements. A significant time will be needed for the development and testing of such fuels before they can be used confidently in fast reactors.

The physics of reactors using fuel that does not contain U-238 has some specific characteristics. The main ones are as follows [2]:

- a large (negative) value for the sodium void reactivity effect (SVRE) of 2 - 3 % $\Delta k/k$;
- a significant decrease in the Doppler effect; and
- a high non-uniformity in the power distribution.

The first feature provides a high degree of reactor safety beyond the normal design accident scenario. The second characteristic is a disadvantage that can be mitigated by the introduction of resonance absorbers into the matrix containing the fuel. Finally, the third characteristic which is also a disadvantage can be compensated by applying special reloading procedures with subassembly interchanging from the periphery to the core centre.

The benefit of using fast reactors in an overall nuclear power system including thermal reactors is not due, as many think, to their ability to burn plutonium, but by their ability to burn minor actinides. In fast reactors, the minor actinides can be fissioned by high-energy neutrons, i.e., they can be used as a nuclear fuel. The simplest way to burn the minor actinides is to mix them with bulk fuel. However, the addition of minor actinides to normal fuel leads to a noticeable increase in the SVRE value. This is the reason why minor actinide burning can be accomplished using only cores with an increased fuel enrichment. Another way of burning the minor actinides involves their use in special subassemblies (SAs) at a high concentration in an inert matrix.

Since curium has a high neutron capture rate and form high level waste isotopes, it is important to separate americium from neptunium so as to reduce curium production during burning. This separation requires the creation of a special production line which includes all of the steps from fuel pin production to reprocessing. This line will require a smaller scale than existing production capabilities. Computational studies show that introduction of minor actinides heterogeneously into the core does not solve the SVRE problem. From this standpoint, the best option is to put specialised subassemblies into radial blankets.

Even more intensive burning of the minor actinides is possible in a specialised fast reactor (that is one with an even more specialised core). The fuel for such a specialised reactor could contain a considerably increased amount of minor actinides. In this case, it is advantageous to use fuel based on an inert matrix that does not contain U-238 because U-238 acts as a source of plutonium and minor actinides during irradiation. Optimisation studies have shown that such fuel can include up to about 30 to 40% minor actinides in a mixture, for example, with U-235. The parameters limiting fuel design in this case are SVRE, β_{eff} and some others. Several such reactors, each capable of burning 500 kg of minor actinides per year, could service the entire nuclear power system of Russia.

3. Limitations on Fuel Composition of Fast Actinide Burners

The allowable amount of heat release in fresh fuel is determined by capacity for heat removal, and depends on the assembly design. Nevertheless, some average values, based on experience already gained, can be adopted for estimation purposes. For instance, an allowable heat release value of approximately 32 to 35 W per kg of plutonium is used in this paper. Table 1 gives heat release values for two plutonium compositions, that differ primarily by their contents of Pu-238 (and Pu-239). Composition A contains Pu^{238}/ Pu^{239}/ Pu^{240}/ Pu^{241}/ Pu^{242} isotopic percentages of 2/55/25/15/3 %, respectively. Composition B contains Pu^{238}/ Pu^{239}/ Pu^{240}/ Pu^{241}/ Pu^{242} isotopic percentages of 5/52/25/15/3 %, respectively.

TABLE 1. Heat release per kg of plutonium for different plutonium isotopic compositions (W/kg)

	Composition A	Composition B
Pu-238	11.2	28.0
Pu-239	1.05	0.99
Pu-240	1.73	1.73
Pu-241	0.65	0.65
Pu-242	0.003	0.003
Total	14.63	31.35

Next consider the change in heat generation rate with the addition of specific minor actinides into the plutonium of composition A. The results for neptunium, americium and curium additions are presented in Tables 2 to 4.

TABLE 2. Heat release (W/Kg) from mixture vs neptunium fraction

Np Fraction	0 %	1 %	10 %	100 %
Np-237	0	$2.16*10^{-4}$	$2.16*10^{-3}$	$2.16*10^{-2}$
Total Pu and Np	14.63	14.47	13.16	$2.16*10^{-2}$

TABLE 3. Heat release (W/Kg) from mixture vs americium fraction

Am Fraction	1 %	10 %	30 %	50 %	100 %
Am-241	0.72	7.2	21.6	36.0	72
Am-242m	0.065	0.65	1.95	3.25	6.5
Am-243	0.023	0.23	0.69	1.15	2.3
Total Pu and Am	15.28	21.24	34.47	47.7	80.8

Americium composition: $Am^{241}/Am^{242m}/Am^{243}$: 0.63/0.021/0.342

TABLE 4. Heat release (W/Kg) from mixture vs curium fraction

Cm fraction	1 %	10 %	30 %	50 %	100 %
Cm-243	0.49	4.88	14.64	24.4	48.8
Cm-244	27.47	274.7	827.1	1373	2747
Total Pu and Cm[*]	42.72	292.7	852.0	1405	2796

[*] Curium composition: Cm^{243}/Cm^{244} equals 0.026/0.977;
Cm^{242} was not taken into account because its half-life is 162 days.

The results just presented lead to the following conclusions:

- no problems were identified with heat release due to the addition of neptunium to the fuel;
- introduction of americium up to 30 % is allowable from a heat generation standpoint;
- introduction of curium (containing mainly the Cm-244 isotope) is limited to less than 1 % (provided that Cm-242 has completely decayed).

Consider in the same sequence the dose rates from assemblies containing mixed (uranium-plutonium) oxide and different amounts of minor actinides. The computations were performed for assemblies of the BN-800 type, with a fuel enrichment of 30 % plutonium based on composition A plutonium and a storage time for the fresh fuel of 3 years. The results are listed in Tables 5 to 7.

TABLE 5. Surface equivalent dose rates (μ rem/s) for different contents[*] of Np^{237}

Np fraction	0 %	0.7 %	10 %	100 % [*]
γ's from U, Pu and Am^{241} (disintegration of Pu^{241})	90	89.4	81	0
neutrons from U, Pu and Am^{241}	110	109	99	0
γ's from Np^{237} and a product of its disintegration, Pa^{233}	0	10	143	1430
Total	200	208.4	323	1430

[*] Content of neptunium in the fuel is given by Np/(Np+Pu+U) in %.

TABLE 6. Surface equivalent dose rates (μ rem/s) for different contents[*] of the americium (isotopic composition of americium is as given earlier at the bottom of Table 3)

	0 %	3 %	30 %	100 % [*]
γ's from U and Pu	90	87.3	63	0
neutrons from U and Pu	110	106	77	0
γ's and neutrons from U and Pu	200	193.3	140	0
γ's from Am	0	1677	16770	55900
neutrons from Am	0	48	480	1600
Total	200	1918	17390	57500

[*] Content of americium in the fuel is given by Am/(Am+pu+U) in %.

TABLE 7. Surface equivalent dose rates (μ rem/s) for different contents[*] of curium (the isotopic composition of curium is as given earlier at the bottom of Table 4)

	0 %	1 %	10 %	50 %	100 %*
By γ-radiation from U, Pu	90	89	81	45	0
By neutron radiation from U, Pu	110	109	99	55	0
By γ- and neutron radiation from U, Pu	200	198	180	100	0
By γ-radiation from Cm (mainly from Cm^{243})	0	5.6×10^3	5.6×10^4	2.8×10^5	5.6×10^5
By neutron radiation from Cm (mainly from Cm^{244})	0	7.5×10^4	7.5×10^5	3.75×10^6	7.5×10^6
Sum value	200	8×10^4	8×10^5	4×10^6	8×10^6

[*] Content of curium in the fuel is given by Cm/(Cm+Pu+U) in %.

The analysis carried out for the surface dose rates, depending on the fuel composition, allows the following conclusions to be drawn. The addition of neptunium to oxide fuel up to a level of 0.7 % (this level corresponds to the case for the steady-state fuel composition after multiple recycles in BN-reactors) does not significantly impact the dose rate related considerations for fuel handling. An increase in the quantity of neptunium requires an increase in the required shielding (steel) of about $\sim$5 cm. The addition of americium to the fuel is possible up to a level of 3 % from the radiation safety standpoint. In this case, the thickness of the radiation shielding needs to be increased by $\sim$6 cm (steel). Adding even a small quantity of curium ($\sim$1 %) to the fuel will significantly complicate fuel handling. Special considerations (remote handling) are necessary in this case.

Initial computational results have been obtained for the fuel cycle characteristics of nuclear power systems consisting of thermal and fast reactors corresponding to the use of different types of fast reactor cores of the BN-800 type. In such systems, a great quantity of plutonium and minor actinides is accumulated as a result of thermal reactor operation. Thermal reactors of the VVER-1000 type were considered both with uranium (only) loadings and with 30 % plutonium loadings. The latter case imposes additional requirements on the fast reactors, i.e., besides completely using the plutonium and actinides in the VVER spent fuel, they need to also supply plutonium for the next thermal reactor loading. Thus, in such a system, the fast reactors play the role of both "burner" and "breeder". The nuclear power systems studied are shown schematically in Fig.1. The annual production of actinides by the thermal reactors using different types of fuels are given in Table 8. The fast reactor design considered is based on the BN-800 type with different cores distinguished by fuel volume fraction, and consequently, by efficiency for actinide burning. The major characteristics of such cores are presented in Table 9.

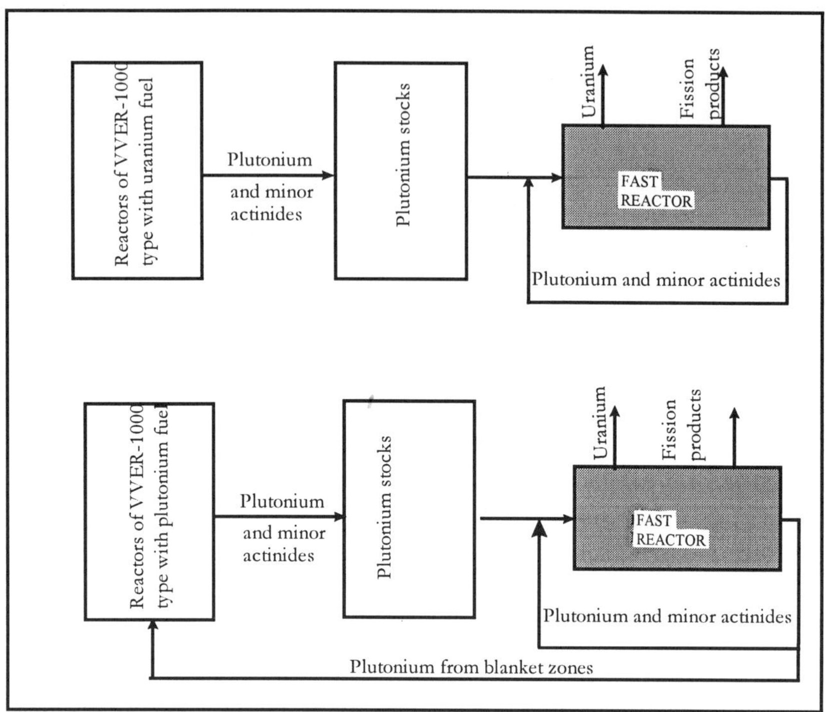

Fig. 1. Principal Schemes for Nuclear Energy Systems Using Both Thermal and Fast Reactors.

TABLE 8. Actinide quantities in the spent fuel of various thermal reactors (kg/year per GW(e))

Isotope	Uranium fuel	Uranium-plutonium fuel
Total	235	385
Pu^{238}	2.8	6.8
Pu^{239}	121.7	160.1
Pu^{240}	53.2	96.7
Pu^{241}	27.5	56.9
Pu^{242}	12.8	31.8
Np^{237}	7.1	11.2
Am^{241}	6.4	8.0
Am^{242}	0.006	0.03
Am^{243}	2.6	9.1
Cm^{242}	-	0.05
Cm^{244}	1.0	4.0
Cm^{245}	-	0.2

TABLE 9. Major characteristics of fast reactor cores

Model number	Fuel volume fraction	Average loading % Pu	Actinide burning efficiency, kg/TW-hr
1	0.31	34.0	50.8
2	0.29	37.0	56.8
3	0.26	41.6	63.9
4[*]	1.10	100	110

[*] Fuel without U-238 is used.

During different studies, the maximum burnup of the fuel was varied (from 10% in the initial variants up to 20 %). For dynamic investigations of the fuel cycle, the decay (storage) time (T_{st}) for the fast reactor's spent fuel from discharge until its reloading into the core was also varied from 1 to 3 years. In the computational studies, two- and three-dimensional codes (RBR-90 and TRIGEX) were used. The calculations of the changing isotopic content of the fuel were performed using the CARE code. The domestic nuclear data library, BNAB, was used in these calculations.

The computational studies performed have shown that for the nuclear power systems considered, a quasi-equilibrium isotopic content is established for the fuel loaded to the fast reactor. Table 10 presents the fuel loading in the established operational regime for systems with different characteristics. The fuel isotopic compositions (given in Table 10) are established after 16-18 operational cycles for system models 1, 2 and 3 while it took 25 cycles for system model 4 to obtain a quasi-equilibrium content. All the compositions are characterised by an increase in the amount of minor actinides in the fuel. If the initial content of minor actinides in the fuel is 7% , the steady state fuel composition will contain 10-13 % minor actinides if the fuel consists of mixed (uranium-plutonium) oxide fuel with an increased initial enrichment and ~ 20 % minor actinides if the fuel does not contain uranium and consists of an inert matrix of magnesium oxide.

In order to assess the fuel's radiation characteristics, the activity and heat generation from one subassembly was calculated. The calculated results are given in Table 11. The results presented show that the specific heat generation increases 3 to 3.5 times when compared to fresh fuel and significantly exceeds the allowable value. The main contributors to the total heat release rate are the curium isotopes (70 %). An increase in the fuel decay (storage) time after irradiation to 3 years achieves a decrease in the fuel heat release by 10 % for the oxide fuel and by 200 % for the fuel without uranium.

188

TABLE 10.   Quasi-equilibrium isotopic compositions of (loaded) fast reactor fuel (kg/ton)

Isotope	Model 1				Model 2			
	burn-out 10 %		burn-out 20 %		burn-out 10 %		burn-out 20 %	
	T_{st}=1yr	T_{st}=3yr	T_{st}=1yr	T_{st}=3yr	T_{st}=1yr	T_{st}=3yr	T_{st}=1yr	T_{st}=3yr
U^{235}	2.5	2.5	2.3	2.3	2.4	2.3	2.1	2.1
U^{238}	621.7	614.5	565.5	559.9	584.1	575.9	521.4	516.4
Pu^{238}	10.8	13.0	11.9	13.3	12.3	14.7	13.5	15.0
Pu^{239}	143.2	141.9	158.9	160.3	151.4	153.8	170.1	171.6
Pu^{240}	124.2	126.2	139.2	140.4	137.6	140.3	153.6	155.5
Pu^{241}	26.6	23.5	33.6	31.2	30.0	26.6	38.1	35.6
Pu^{242}	33.4	33.6	41.2	41.4	38.6	38.4	47.3	47.4
Np^{237}	5.2	5.4	6.9	7.0	6.0	6.3	8.0	8.2
Am^{241}	11.0	15.8	12.1	17.1	12.4	17.9	13.8	17.6
Am^{242}	0.6	0.8	0.6	0.7	0.7	0.9	0.7	0.8
Am^{243}	12.3	12.5	15.8	15.8	14.3	14.3	18.0	18.2
Cm^{242}	0.2	0.02	0.2	0.02	0.2	0.02	0.2	0.02
Cm^{244}	6.7	5.5	9.1	8.1	7.8	6.8	10.4	9.1
Cm^{245}	1.8	1.5	2.4	2.2	2.1	1.8	2.7	2.5

Isotope	Model 3				Model 4	
	burn-out 10 %		burn-out 20 %		burn-out 50 %	
	T_{st}=1yr	T_{st}=3yr	T_{st}=1yr	T_{st}=3yr	T_{st}=1yr	T_{st}=3yr
U^{235}	2.1	2.1	1.8	1.8	—	—
U^{238}	530.1	520.4	452.0	444.3	—	—
Pu^{238}	14.6	17.3	16.0	17.8	46.6	53.5
Pu^{239}	164.0	166.7	187.9	190.4	249.0	250.1
Pu^{240}	156.9	159.6	177.1	179.1	358.7	358.3
Pu^{241}	35.6	31.7	45.2	42.5	80.2	67.9
Pu^{242}	45.3	45.8	55.8	56.7	118.6	119.0
Np^{237}	7.3	7.7	9.7	10.0	17.1	17.8
Am^{241}	14.7	20.9	16.4	20.8	47.2	62.1
Am^{242}	0.8	1.1	0.8	0.9	3.0	3.5
Am^{243}	16.7	16.8	21.8	21.6	46.6	43.7
Cm^{242}	0.3	0.03	0.2	0.03	0.6	0.06
Cm^{244}	9.3	8.0	12.1	10.8	28.3	20.1
Cm^{245}	2.3	2.0	3.1	2.8	5.3	4.0

TABLE 11. Heat release rates of fresh subassemblies with quasi-equilibrium fuel compositions (W/kg Pu)

	Burn-out 10 %			Burn-out 20 %		
	fresh	established		fresh	established	
		T_{st}=1yr	T_{st}=3yr		T_{st}=1yr	T_{st}=3yr
Model 1	24.2	76.6	67.9	24.2	83.7	76.0
Model 2	24.1	81.3	74.5	24.1	88.9	80.5
Model 3	24.1	85.3	75.7	24.1	90.1	81.2
Model 4	23.7	171.7	96.90	—	—	—

One of the main conclusions from the consideration of the nuclear power systems, consisting of VVERs using uranium fuel and BN-800 reactors, is the importance of separating the curium isotopes from the fuel downloaded from the fast reactors. Similar studies have been carried out for the fuel cycles of nuclear power systems using VVERs with a 30 % loading of plutonium. The computational results show that the established (quasi-equilibrium) fraction of minor actinides in the plutonium is 15 to 18 % which is 1.5 times higher than that for a system with uranium fuelled VVERs. The corresponding heat release rates for fresh subassemblies based on the quasi-equilibrium fuel compositions for this case are presented in Table 12.

TABLE 12. Heat release from fresh subassemblies for established fuel compositions (W/kg)

	Burn-out 10 %			Burn-out 20 %		
	fresh	established		fresh	established	
		T_{st}=1yr	T_{st}=3yr		T_{st}=1yr	T_{st}=3yr
Model 1	45.1	105.3	92.9	45.1	115.1	97.6
Model 2	45.1	114.6	103.2	45.1	125.2	109.7
Model 3	45.1	125.6	111.7	45.1	137.2	115.4
Model 4	44.2	290.6	165.4	—	—	—

As can be seen from Tables 11 and 12, use of the plutonium downloaded from VVERs with an initial 30 % plutonium loading leads to a doubling of the heat release rate in the fresh fuel. This increase is mainly due to the large fractions of Am-243 and Cm-244 in this fuel. Consequently, a decrease in the curium content in the fresh subassemblies is needed for handling purposes. Thus, when formulating closed fuel cycles for nuclear power systems employing fast (burner) reactors and thermal reactors, it is necessary to also separate the curium isotopes with the aim of decreasing the heat release rate and activity of the fresh subassemblies with the quasi-equilibrium composition.

From the fuel cycle studies, it is possible to determine the approximate number of VVER-1000 reactors whose wastes could be treated by one fast reactor of the BN-800 type. The results of these calculations are given in Table 13. One reactor of the BN-800

TABLE 13. Number of VVER-1000 reactors whose plutonium could be utilized by one BN-800 reactor.

	Burnup 10 %		Burnup 20 %	
	T_{st}=1yr	T_{st}=3yr	T_{st}=1yr	T_{st}=3yr
Model 1	1.35 (0.86) *	1.32 (0.85)	1.53 (0.97)	1.54 (0.97)
Model 2	1.53 (1.00)	1.46 (0.99)	1.67 (1.06)	1.68 (1.06)
Model 3	1.63 (1.06)	1.63 (1.05)	1.85 (1.17)	1.86 (1.17)
Model 4	2.6 (2.6)	2.6 (2.6)		

* The values in parentheses denote the number of VVERs assuming their operation with a 30 % plutonium loading.

type using mixed oxide fuel with an increased enrichment can utilise the plutonium and minor actinides produced by approximately 1.5 to 1.8 reactors of the VVER-1000 type using uranium loadings and ~1.0 reactor of the VVER-1000 type using a 30% loading of plutonium fuel. A BN-800 type reactor using fuel that does not contain U-238 can utilise the plutonium and minor actinides, produced by ~2.6 reactors of the VVER-1000 type for either type of VVER fuel loading.

4. Conclusions

The studies performed to determine the characteristics of nuclear power systems consisting of thermal reactors of the VVER-1000 type and fast (burner) reactors of the BN-800 type lead to the following conclusions. In combining VVER reactors and fast (burner) reactors in a system with repeated fuel recycling, the quantity of minor actinides in the fuel increases monotonically until reaching an equilibrium after 18 to 25 cycles, depending on the core type used for the fast reactors. For the initial conditions adopted in this work, the minor actinide level increases from 7 % (relating to the total quantity of plutonium) to 10 to 13 % for oxide cores and to 20 % for cores with fuel that does not contain U-238. Values of the heat release rate and external dose rate from fresh fuel with these levels of minor actinides can meet modern safety requirements only on the condition that curium is separated from the irradiated fuel. The treatment of curium from a high level waste standpoint must be considered separately. Following this scenario, a fast burner reactor of the BN-800 type can utilise the plutonium and minor actinides from 1 to 2.6 reactors of the VVER-1000 type, depending on the types of core used by the BN-type reactors. Thus, the possibility exists for creation of a nuclear power system that consists of thermal and fast reactors operating jointly with a (nearly) closed fuel cycle in which all of the major actinides (except curium) are utilised.

It is worth noting that the main assumptions and data used in this work may be subjected to change. However, the authors believe that the principal conclusions would nevertheless remain the demonstration of the possibility for (nearly) complete actinide utilisation in a closed nuclear power system using fast reactors.

References

1. Krivitski I.Y. and Matveev V.I. (1995) Development of fast reactor cores for weapons-grade plutonium utilization, E.R. Merz, C.E. Walter and G.M. Pshakin (eds.) *Mixed Oxide Fuel (MOX) Exploitation and Destruction in Power Reactors*, NATO ASI Series.
2. Krivitski I.Y., Matveev V.I. and Burievski I.V. (1995) Evolution of physical concepts of fast reactor cores for effective actinide consumption, *Proceedings of the Third International CAPRA Seminar,* Lancaster, UK.
3. Krivitski I.Y., Byburin G.G., Ivanov A.P., Matveev V.I., Matveeva E.V. (1995) Pu burning in fast reactor cores using unconventional fuel without U^{238}, *Unconventional options for plutonium disposition* IAEA-TECDOC-840.

PLUTONIUM-FUELED LMFRs: PROBLEMS OF DESIGN OPTIMIZATION FOR SELF-PROTECTION

A.M. KUZMIN
V.S. OKUNEV
Moscow State Engineering Physics Institute (MEPhI)
31, Kashirskoe shosse,
Moscow, 115409, RUSSIA

1. Introduction

The design of the next generation of nuclear powerplants requires improvements in reactor safety. Safety improvement is also a problem of great importance for currently operating units. Nuclear powerplant safety may be guaranteed by successful approaches to the following issues: 1) self-protection of the core by natural mechanisms during accidents (ATWS (Anticipated Transient Without Scram) accidents are of primary importance), 2) civilian and weapons-grade plutonium utilization, 3) treatment and disposal of radioactive wastes and 4) assurance against proliferation of nuclear materials.

2. LMFR Layout for Plutonium Utilization

The LMFR (Liquid Metal Fast Reactor) concept for plutonium utilization and treatment of minor actinides is most interesting and simple. In this concept, the reactor core employs a higher than traditional fissile fuel enrichment and a steel blanket. Some characteristics of the type of LMFR considered in this paper are presented below. It is based on the BN-800 design concept. The thermal power of the reactor is 2100 MW, the electrical power is 800 MW and the sodium inlet temperature is 627 K. The external dimensions of the core (with blanket) correspond to those of the BN-800 reactor. The other core parameters for the reactors considered here are different for each case. They were obtained by optimizing the cores for different goals. In this study, it was assumed

T. A. Parish et al. (eds.),
Safety Issues Associated with Plutonium Involvement in the Nuclear Fuel Cycle, 193–198.
© 1999 *Kluwer Academic Publishers. Printed in the Netherlands.*

that weapons-grade plutonium consists of ^{239}Pu (93%) and ^{240}Pu (7%) and that (civilian) reactor-grade plutonium (from LWRs) consists of ^{238}Pu (2%), ^{239}Pu (61%), ^{240}Pu (24%), ^{241}Pu (10%) and ^{242}Pu (3%) [1].

Currently, there are several different types of LMFR cores under development for plutonium utilization and minor actinide burning. It should be noted, that from a purely physics standpoint, it is preferable to utilize either reactor-grade or weapons-grade plutonium in LMFRs where it can be used more efficiently to produce energy than in LWRs (Light Water Reactors).

3. Problems of Design

Design of LMFR cores for plutonium utilization or minor actinide burning requires development of a methodology for optimizing the designs for each mission. The requirements for the physical layout of the cores were formulated in the optimization procedure as constraints imposed on some functionals that describe a design's performance. These functionals are: 1) safety-related functionals (maximum temperature, power, pressure etc. in abnormal situations), 2) reliability-related functionals (used to characterize the availability, longevity, repair and maintenance needs of the major nuclear power plant components) and fuel-related functionals (used to characterize the fissile fuel requirements, waste production, etc). The functional(s) characterizing the economic efficiency (cost) can be considered as the goal (or objective) functional(s).

Two different types of problems related to the design of self-protected cores were considered. The first type is a problem of design optimization. The second type of problem is to determine the domain that identifies inherently safe core designs.

The optimization problem can be formulated in the following way: To determine the vector of control parameters (or control vector), $\mathbf{u}_o$, such that the goal functional, F_0 (or goal functionals), is an extremum while simultaneously satisfying the constraints from the reliability-related functional(s), the fuel-related functional(s) and the safety-related functionals (maximum temperatures, power etc.) during ATWS. The control vector consists of parameters related to the configuration of the core layout: sizes of fuel zones and characteristics of the lattice, fuel enrichment, flow rate in normal operation, characteristics and fraction of some materials: fuel, coolant, structures, control rods, scram rods etc.). The determination of the parameters in the control vector are important in the initial phase of any study to formulate a new reactor core design. In the initial phase of such a study, some functionals may be considered as goal functionals, or alternatively, can be imposed as constraint functionals [2]. The without-goal-functional problem for defining the domain of self-protected core layouts in the control vector space can be considered as a particular case of such problems. Possible methods for solution of without-goal-functional problems have already been developed by the authors and were presented in [3].

The problems of the second type, ie., to determine the domain that identifies acceptable inherently safe core designs, are based on determination of the domain for core inherent safety. This domain for LMFR cores is characterized by maximal permitted perturbations of reactivity, flow rate and inlet coolant temperature. These perturbations don't require active engineered safety features to assure that the safety mechanisms operate. In this second type of problem, the amplitudes of perturbations, the rate or duration of different perturbations, and the comparative delay times for different transients or perturbations can be considered as parameters in either the goal functional or in the constraint functionals.

The solution of full optimization or without-goal-functional optimization problems can also be very interesting. In this problem, the unexpanded control vector contains the components of control vectors for the problems of the first and second types.

4. Accounting for Uncertainties in Initial Data

Uncertainties in the initial data are always present in practical problems of core design. The solutions of a problem with uncertain initial data are within the domain of the optimal control vector.

The most interesting case is to take into account uncertainties of initial data with known probabilistic characteristics in the first and second types of problems and to combine problems with constraints imposed on the safety-related functionals. Such uncertainties are the typical ones for the operational characteristics of the core. These are as follows: coastdown time, time of reactivity insertion, time for coolant inlet temperature changes, time for coolant circulation in the primary and secondary coolant circuits (loops), reactivity margins, characteristics of the emergency core cooling system and so on. The rather large variations in some of this data must be considered in core design.

5. Some Results for LMFR Design

As an example, the solution of an optimization problem are presented for an LMFR employing civilian plutonium nitride fuel for the case of constraints imposed upon safety-related functionals characterizing the LOF WS (Loss of Feedwater Without Scram) and TOP WS (Transient Over Power Without Scram) combination. It is also assumed that the initial data for ATWS are uncertain.

The results for the void reactivity effect (VRE) minimization are presented in Figure 1. In this figure, F_0 denotes the goal functional (value of the VRE), t_c is the coastdown time in LOF WS, "dρ" specifies the value of the external reactivity insertion in the TOP WS, "ρ" denotes the possibilities for the optimized solution, ie., the domain of optimal F_0. The best value of the VRE is reached at the point "O" in Figure 1 (optimistic scenario of ATWS), the worst value of the VRE is reached at the point "P" of Figure 1 (pessimistic scenario of ATWS).

In some cases, the optimistic and pessimistic scenarios do not lead to unique solutions. In these cases, the domain of optimal F_0 can have a complex form. However, these cases occur very seldomly.

196

The results for the problems of the second type are presented in Figure 2. In this figure, "α" is the domain of cores with inherent safety, "δG" is the variation of flow rate ("δG">0 - LOF WS, "δG" < 0 - OVC WS), and "δρ" is the reactivity insertion. These results were obtained for the LMFR core, corresponding to the point "P" in Figure 1. The vertical shading shows the effects of uncertainties in the initial data. In cases in which the safety-related functionals are a maximum for steady-state operation (after a transient has died away), the border of the "α"-domain is defined unambiguously.

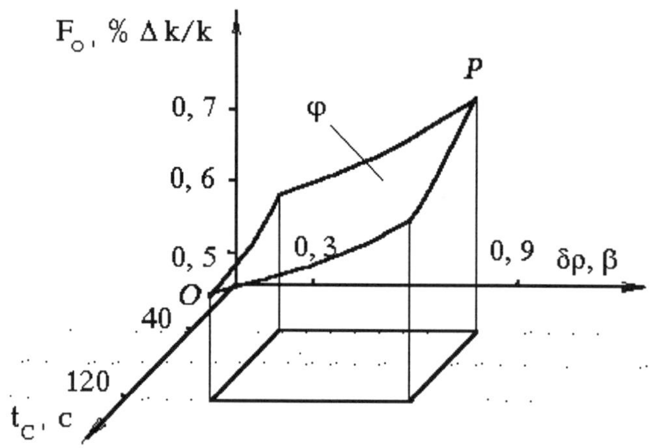

Fig. 1. Results of VRE minimization.

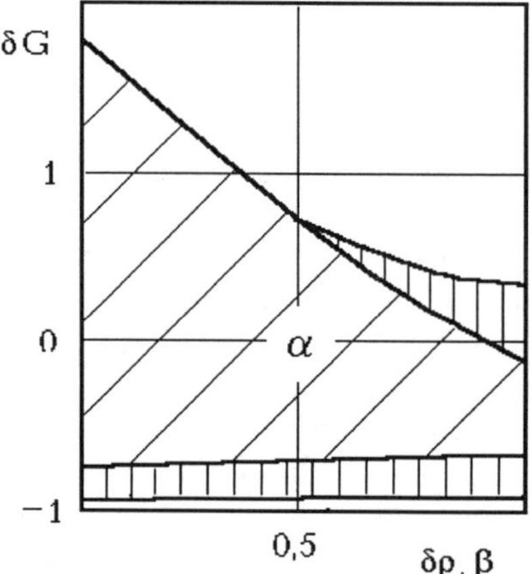

Fig. 2. Domain of core inherent safety in ("δρ", "δG") co-ordinates

6. Analysis of Self-protected Cores

Additional serious accidents for both optimal LMFR cores (corresponding to the points "O" and "P" of Figure 1) were investigated. These are LOF WS, TOP WS, OVC WS, LOHS WS and their combinations.

It should be noted that more serious accidents are realized when LOF WS, TOP WS and OVC WS are combined and OVC WS starts later than LOF WS and TOP WS. The combination with a simultaneous beginning for LOF WS, TOP WS, OVC WS and LOHS WS is less serious because some of the transients neutralize one another (for example, LOHS WS and OVC WS are self-neutralized). The core designs obtained as a result of solving the optimization problems under conditions of optimistic and pessimistic scenarios of ATWS are quite safe. Even in the more serious accident cases, the coolant doesn't boil, the cladding isn't destroyed, and the fuel doesn't melt.

A comparative analysis has been performed for LMFR cores, that are self-protected against severe accidents, using both (civilian) reactor-grade and weapons-grade plutonium. LMFRs of the BN-type were considered. Figure 3 shows the time dependence of the maximum fuel temperature during a LOF WS. The time dependence of the maximum fuel temperature for the designs considered are determined largely by

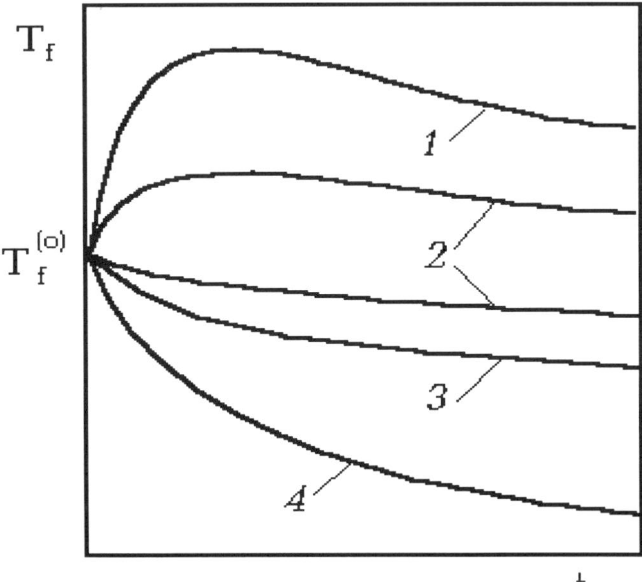

Fig. 3. Time-dependence of maximum fuel temperature in LOF WS for BN-600 and BN-800-types of LMFRs for plutonium utilization.(1. Metal fuel (and metal with Zr), 2. Nitride or carbide, 3. Mononitride or monocarbide, 4. Oxide)

the type of fuel. These studies affirmed the assertion of Hammel and Okrent about the role of Doppler reactivity feedback in transients when the flow rate is decreasing. The

198

time dependence of the maximum fuel temperature in low-powered LMFRs with different plutonium fuels corresponds to curve number 1 in Figure 3. This plot is characteristic of low-powered LMFRs. Thus, the requirement to design self-protected reactor cores can be formulated using constraints imposed on the safety functionals in abnormal situations (including ATWS). Satisfaction of the constraints imposed by other functionals, for example, reliability-related functionals, characterizing the availability, longevity, reparability and maintainance needs of the major nuclear powerplant components in normal operation, do not necessarily guarantee the self-protection of the reactor core during severe accidents.

7. Alternative Fuel Cycle

The possibility of introducing thorium into the LMFR fuel cycle while allowing for the use of plutonium was also considered. Two different types of LMFRs with thorium fuel were considered. The first of was based on the thorium-uranium fuel cycle and the second was based on a mixed thorium-uranium-plutonium fuel cycle. The LMFRs with mixed fuel cycles were considered for both reactor-grade and weapons-grade plutonium utilization. Mixtures of ^{233}U and ^{238}U with low ^{233}U fractions have less value as potential weapons material.

The Th-^{233}U fuel is characterized by a high melting temperature, a small thermal expansion coefficient and a high thermal conductivity. Cores using thorium fuel exhibit better self-protection during TOP WS and OVC WS than traditional fast reactors fuelled with U-Pu. However, the advantages of LMFRs using thorium fuel are very slight. LMFR fuel, based on the thorium cycle, is expected to be much more expensive than conventional fuel.

8. Conclusions

Fuels based on reactor-grade or weapons-grade plutonium can be used in LMFRs. However, the operational characteristics of the reactor cores during such utilization have much smaller safety margins than those of traditional LMFRs. Self-protected cores for LMFRs using the Th-^{233}U or thorium-uranium-plutonium (with either reactor-grade or weapons-grade plutonium) fuel cycles aren't much better than LMFR cores fuelled with Pu.

References

1. Rahn,F., Ademantiades,A., Kenton,J., Brawn,C. (1984) *A Guide to Nuclear Power Technology,* New-York.
2. Kuzmin,A.M. and Okunev,V.S. (1996) Self-Defence of LMFRs Core Design: Computer Codes and Preliminary Results, *International Conference on the Physics of Reactors "PHYSOR-96"* **3**, J171-J178.
3. Kuzmin,A.M. and Okunev,V.S. (1996) The Methods for Research of Without-Goal-Functional Problems for Definition of the Domain of Safe Core Layouts, *Atomic Energy,* **81/3**, 170-178.

EQUILIBRIUM, PROLIFERATION RESISTANT, CLOSED FUEL CYCLES FOR LWRs

A.N. CHMELEV
G.G. KOULIKOV
V.B. GLEBOV
V.A. APSE
Moscow State Engineering Physics Institute (Technical University)
Department of Theoretical and Experimental Reactor Physics
Russia 115409, Moscow, Kashirskoe shosse, 31

1. Introduction

It is known that a U-Pu fuel cycle including reprocessing and recycling of nuclear fuel evokes contradictory opinions with respect to the potential risk of Pu proliferation. This argument can be decomposed as follows:

- although Pu extracted from the spent fuel of power reactors (for example, LWRs of the PWR, BWR, or VVER types) is not the best material for nuclear weapons, it can still be used in nuclear explosive devices of moderate energy release [1];
- in the case of a closed fuel cycle, recycled Pu will be separated at chmical processing facilities thereby increasing the probability of this material being used for illegal aims (diversion, theft).

Under these assumptions and in the absence of an internationally coordinated plan for the utilization and/or disposal of spent fuel, the leading nuclear countries were forced to undertake steps directed to strengthen their nonproliferation efforts (IAEA safeguards program and Euratom's embargo on the export of spent fuel reprocessing technology). In addition, several countries, including the U.S., stopped the deployment of breeder reactors, which were intended for operation with closed fuel cycles, and focused instead on once-through fuel cycles. From another perspective, the social demand for solving the problem of how to disposition excess fissile materials (plutonium, most of all) which have both civil and military origin, has stimulated research into Pu utilization in MOX fuel. Concurrently, studies of advanced fuel cycles protected against uncontrolled proliferation of fissile materials have been initiated. This paper is devoted to

199

T. A. Parish et al. (eds.),
Safety Issues Associated with Plutonium Involvement in the Nuclear Fuel Cycle, 199–212.
© 1999 *Kluwer Academic Publishers. Printed in the Netherlands.*

the description of some approaches to Pu protection through denaturing, i.e. by purposely altering the Pu isotopic composition.

One of the ways to improve Pu protection against uncontrolled proliferation is to mix it with the isotope Pu-238 (with a half life of 87 years) which is characterized
1. by intensive heat generation, ie., a specific heat generation rate of 570 W/kg and
2. by the generation of spontaneous neutrons [1-6]. Both these factors can be very important contributors to Pu protection. It is proposed to produce Np-237, Pu-238 and Pu-239 in blankets of controlled thermonuclear reactors (CTRs) for feeding, and thus, denaturing the MOX-fuel of LWRs. The proposed fuel cycle involves ternary fuel compositions and may be called a Np-U-Pu fuel cycle.

An equilibrium proliferation resistant Th-U fuel cycle has also been analyzed. The heat-spiking concept is used to enhance proliferation resistance of this fuel cycle. Fusion facilities with thorium-containing blankets can generate Pa-231 and U-232. It has been shown that a high content of U-232 in the uranium for the equilibrium Th-U fuel cycle of LWRs can be achieved and maintained. Such denaturing can achieve a high level of protection for the U-233 against uncontrolled proliferation. It has also been found that this fuel cycle is characterized by an important increase in the fuel burnup and moderate reactivity changes.

2. On Enhancement of the (LWR) MOX Fuel Cycle's Proliferation Resistance by Plutonium Denaturing

2.1. MAIN ASSUMPTIONS

1. The equilibrium isotope vectors are obtained for MOX fuel assemblies circulating between LWR, spent fuel reprocessing and fuel manufacturing facilities.

2. The fuel feed includes the isotopes, Np-237, Pu-238 and Pu-239. The fuel feed is produced in CTR blankets in which depleted uranium is irradiated by fast (14 MeV) fusion neutrons which initiate threshold (n,2n) and (n,3n) reactions in U-238. Depending on the uranium layer thickness and the moderating properties of the blanket, various correlations between the amounts of generated Np-237, Pu-238 and Pu-239 can be obtained.

3. The effect of the introduction of Np-237 and Pu-238 on the LWR's safety characteristics was not considered in this study.

2.2. Pu PROTECTION OF LWR FUEL VERSUS Pu PROTECTION OF LWR FEED

Using the GETERA code [7] for calculations of fuel burn-up, the Pu isotopic compositions of MOX fueled PWRs were determined at beginning and end of life. The Pu-238 fraction in plutonium was adopted as an index to indicate Pu protection against uncontrolled proliferation. This means that the impact of higher Pu isotopes on the neutronics of the chain reaction in implosion-based nuclear explosives using plutonium was not taken into account.

The fuel unloaded from a PWR may be considered to consist of two parts. The first part (due to the "fertile" part of fuel) is composed of the residual U-238 and Pu isotopes produced by U-238. The second part (due to the "fissile" part of fuel) is composed of the Np-237, Pu-238 and other Pu isotopes produced entirely by the fuel's "fissile" portion. Maintaining the Pu-238 fraction of the plutonium in PWR fuel depends on Pu-238 production from both the "fertile" part and the "fissile" part of the loaded fuel and the Np-237 sustained in the "fissile" part of the loaded fuel.

Pu-238 contents in the loaded and unloaded fuel for multi-cycle operation with various Np-237 fractions in the loaded fuel are presented in Fig.1. The region situated under line B in Fig. 1 indicates a regime in which the (loaded) feed fuel is better protected than the discharged fuel. Equally, the region above line B represents a region in which the discharged fuel is better protected than the feed fuel. The curves in Fig. 1 characterize the correlation between Pu protection levels of the feed and discharged fuel when the "feed" part of the fuel includes Np-237 as an additive along with the Pu. Based on these data, it is possible to select an appropriate operational regime for the new nuclear fuel cycle.

Various combinations of the feed compositions, i.e. fractions of Pu-238 and Np-237, are able to attain the same level of proliferation protection in the (discharged) Pu. For example, a 32% (Pu-238) level of protection in the discharged Pu can be achieved in cases of feed fuel containing (0% Np-237 and 52% Pu-238) or (20% Np-237 and 43% Pu-238) or (40% Np-237 and 32% Pu-238). The latter option corresponds to an equal level of Pu protection both in the discharged and feed Pu. The line denoted by "S" in Fig. 1, that connects the right ends of the curves shown there, may be regarded as an "ultimate option" for the Np-U-Pu cycle considered here. The points of this line correspond to a limiting option for the Np-U-Pu fuel cycle in which U-238 is absent from the fuel composition, and its "fertile" function is entirely passed to Pu-238 and Np-237. In effect, this cycle becomes a Np-Pu fuel cycle. In this case, the highest possible (discharged) Pu protection level (65% Pu-238) is reached with feed Pu protection of 90% Pu-238. It is well known, that IAEA safeguards do not apply to Pu containing 80% Pu-238 or more [8-10].

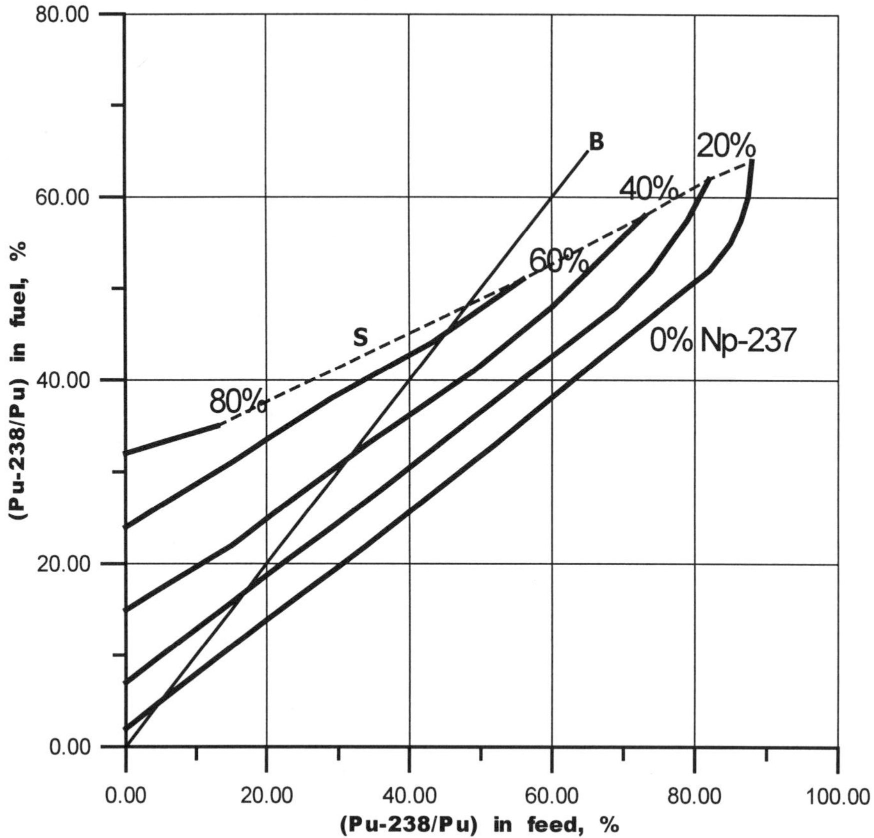

Fig 1. Pu-238/Pu content (proliferation resistance) of Pu in discharged fuel as a function of proliferation resistance of Pu in feed fuel for various Np-237 contents in the feed fuel. "B" denotes the bisectrix.

Internal heat generation in Pu is considered to be a significant factor for its protection. The rates of internal heat generation for various feed compositions are presented in Table 1. The rates of specific heat generation for weapons-grade plutonium (WGPu) and reactor-grade plutonium (RGPu) are presented as well.

TABLE 1. Decay heat generation (q^{Pu}) and neutron generation by spontaneous fissions (n_{sf}^{Pu}) in LWR fuel with equal Pu protection both in fuel and in feed

Generation	WG Pu	RG Pu	Pu-238/ Pu in both discharged and feed fuel		
			(Np /(Np + Pu) in feed)		
			17% (7%)	33% (15%)	44% (19%)
q^{Pu}, W/kg Pu	2.3	13.	97	186	248
n_{sf}^{Pu}, 10^6(n/sec)/kg Pu	0.06	0.38	0.71	1.06	1.3
q^{fuel}, W/kg feed fuel	/ - /	/ - /	14.9	41.2	99.5
n_{sf}^{Pu}, 10^6(n/sec)/kg feed fuel	/ - /	/ - /	0.11	0.24	0.53
Feed Np/Pu-238/Pu-239, kg/(GWe-y)	/ - /	/ - /	38/82/402	103/194/377	176/318/421

Based on the results shown above, it can be concluded that denatured Pu fuel containing more than 25% Pu-238 is characterized by internal heat generation which exceeds that of RGPu by more than an order of magnitude and that of WGPu by an even larger factor. In addition, denatured Pu fuel is characterized by a higher neutron background due to spontaneous fission. The factors mentioned above enhance Pu protection against its utilization in nuclear explosive devices. The same factors complicate, to a certain degree, the handling procedures for reactor fuel.

Values of specific heat generation and neutron emission rates due to spontaneous fission of loaded MOX fuel for the equilibrium cycle options analyzed are shown in Table 1 as well. For comparison, "dry" technology for handling spent fuel assemblies may be applied if the specific heat generation rate does not exceed 20-35 W/kg fuel. It may be also concluded that Pu-denaturing with Pu-238 is restricted by thermal constraints imposed on permissible specific heat generation in the fuel. The same tendency exists in connection with spontaneous neutron emission. These constraints need to be taken into account in fuel fabrication, fuel assembly manufacturing and transport operations. These complications to the fuel cycle may be considered as "costs" for improving the proliferation resistance of the MOX fuel cycle.

2.3. Pu PROTECTION OF LWR FUEL BY Pu PROTECTION OF FEED WITH THORIUM

The fuel cycle considered above, ie., the Np-U-Pu cycle, requires production of the Pu-238 and Np-237 isotopes. It appears that production of these isotopes in sufficient amounts will not be a simple problem to solve. Hence, decreasing the consumption of these isotopes during PWR operation is a very desirable goal. This goal can be achieved by adding thorium to the fuel, leading to a Th-Np-U-Pu fuel cycle. The motivation for such denaturing is the fact that the presence of U-233 (and its daughter isotopes) in the fuel cycle decreases the required Pu content in the fuel, and thus, decreases the demanded quantities of Pu-238 and Np-237. It is worth noting that the U-233 isotope is protected from a proliferation standpoint because of the presence of U-238 in the fuel. As an illustrative example, equilibrium multi-cycle PWR operation has been analyzed. It was assumed that the fissile uranium content (U-233+U-235) in the uranium fraction of the fuel was equal to 20%. According to the IAEA recommendations [8], uranium with fissile fractions less than 20% is not considered to be a direct-use material for nuclear explosive devices.

Plutonium fuel protection in the Th-Np-U-Pu fuel cycle depends on the protection of the feed Pu in a manner similar to that of the Np-U-Pu fuel cycle (see Fig.1). Some parameters of the Th-Np-U-Pu fuel cycle are presented in Table 2 for equal denaturing in the discharged fuel and in the feed fuel.

One can see that the isotopes from the "thorium" fraction in the fuel (Th-232, U-233, U-234, U-235 and U-236) contribute significantly (21-40%) to the total energy release. This allows for a corresponding decrease in the amount of denatured Pu feed. The lower Pu content in the fuel and, consequently, lower Pu-238 loading make it possible to decrease both the specific heat generation in the fuel and the neutron background due to spontaneous fission. The latter effects may lead to simplifications in the fuel handling procedures as compared to those for the Np-U-Pu fuel cycle.

It should be noted that the uranium fraction in the Th-Np-U-Pu fuel will contain U-232 produced by side nuclear reactions. It is known that, the decay of U-232's daughter isotopes is accompanied by hard γ-radiation and, therefore, the presence of U-232 may be considered an additional factor enhancing proliferation resistance of fuel.

Thus, analyses of equilibrium fuel cycles based on application of Np-U-Pu and Th-Np-U-Pu fuels in LWRs show the following:

- it is possible to maintain a high Pu-238 content in the Pu of the (discharged) fuel for an equilibrium LWR fuel cycle by means of appropriate adjustment of the (Pu-238 and Np-237) composition in the feed fuel;
- the use of a Th-Np-U-Pu fuel cycle makes it possible to decrease denatured Pu consumption in an equilibrium cycle.

TABLE 2. Fuel parameters for an equilibrium Th-Np-U-Pu fuel cycle at the beginning of cycle for equal Pu protection in the discharged fuel and in the feed fuel.

Fuel parameters	Pu - 238/Pu and feed fuel feed)	both discharged (Np /(Np +Pu) in
	33%	44%
	(2.5%)	(10%)
Th content in fuel, % HM	68	41
(U-233+U-235)/U in fuel	20%	20%
Fission fraction of (U-233+U-235)	40%	21%
q^{Pu}, W/kg Pu	186	248
q^{fuel}, W/kg feed fuel	22	75
n_{sf}^{fuel}, 10^6(n/sec)/kg feed fuel	0.13	0.39
Feed Np/Pu-238/Pu-239, kg/(GWe-yr)	13/149/293	83/284/339

3. Equilibrium Th-U Fuel Cycle for LWRs with Additional Protection Against Fissile Material Proliferation

In the traditional understanding of Th-U fuel cycles, fuel containing U-233 together with its daughter isotopes is, in essence, highly enriched uranium. The well-known idea is to dilute this uranium with abundant U-238 to enhance the proliferation resistance of this cycle. However, plutonium accumulated in such a fuel cycle will require proliferation protection measures, ie., its denaturing with Pu-238. At the same time, there is the possibility to upgrade the protection level for U-233 in the Th-U fuel cycle without using U-238 and needing to have plutonium produced as a consequence. The isotope for accomplishing this in the Th-U fuel cycle is U-232. Like Pu-238, this isotope is characterized by intense heat generation caused by radioactive decay (its half life equals 68.9 years and its specific heat generation rate is 740 W/kg without accounting for the decay of its daughter isotopes). It is well known that fissile material is applied in highly enriched form in the pits of nuclear explosive devices. The handling of pits produced uranium containing a significant fraction of U-232, would require provisions for intensive heat removal. This last condition may be considered as a factor hampering the use of this material in weapons applications.

It also needs to be noted that the time-dependent behavior of the heat generation caused by the radioactive decays in the U-232 chain changes significantly during the establishment of the equilibrium state between U-232 and its daughter isotopes [11]. There are six successive α-decays in the U-232 chain ending with stable Pb-208. Therefore, the specific heat generation of U-232 in equilibrium with its daughter isotopes is about 9 times higher than that of Pu-238, and this equilibrium is established after about 10 years. For a sufficiently long external fuel cycle, this circumstance will be a factor that hampers fuel handling, especially in the case of incomplete refining of the fuel.

The hard γ-radiation, accompanying the decay of some isotopes of the U-232 chain (thallium-208, bismuth-212) may be considered as additional protection for the fuel. According to the estimate given in reference [11], the maximum exposure rate at a 1 m distance from a 1 mg point source of U-232 in equilibrium with its decay daughters is equal to 12.9 mR/h. This means that, if a pit, with a total weight in the range of kilograms, contains 1 kg of U-232, then the maximum exposure rate at 1 m will exceed 10 000 R/h. Even if this estimated exposure rate were a factor of 10 too high, it remains nonetheless extremely large and is comparable to the exposure rate of LWR spent fuel assemblies. Presently, according to the recommendations of the IAEA and the US Nuclear Regulatory Commission [12], such a dose rate represents an effective protection factor for the plutonium contained in LWR spent fuel assemblies.

A nuclear power system based on MOX fueled LWRs of the VVER type and supplemented with controlled thermonuclear reactors (CTRs) for producing Np-237, Pu-238 and other heavier Pu isotopes was analyzed in [6]. In that paper, operational parameters were evaluated for a VVER fueled with MOX containing Pu-238, and the Pu-238 fraction in the plutonium of the MOX fuel was considered as the index for measuring plutonium protection. It was demonstrated that the Pu-238 content in the plutonium fraction of the fuel changes relatively slowly with time. The slow change of the plutonium protection level during fuel irradiation in an LWR is considered to be a favorable effect.

The earlier studies have been extended by considering the TH-U fuel cycle, and the present paper presents the results obtained. An equilibrium regime is attained for the Th-U fuel cycle with feed fuel containing isotopes of Pa-231, U-232 and U-233. The equilibrium uranium isotope vectors were evaluated for U-232 contents up to 67%. The U-232 fraction in the uranium was adopted as one way of measuring the denaturing of the uranium. Also, for each fuel option, the rates of heat generation from decay were evaluated. These indices (and the intensity of hard γ-radiation) define the fuel's protection level against use in pits of nuclear explosive devices. Also, the specific heat generation rates of the fuel at the refabrication stage, which complicate fuel handling in the fuel cycle, were also evaluated.

3.1. MAIN ASSUMPTIONS

1. An equilibrium multi-cycle operating regime for the Th-U fuel cycle in LWRs is considered.

2. It is assumed that the feed fuel includes isotopes of Pa-231, U-232 and U-233. Needless to say that burn-up of thorium is also accounted for.

3. The feed fuel is produced in CTR blankets in which Th-232 is irradiated by (fast) fusion neutrons to initiate threshold reactions, (n,2n) and (n,3n) reactions, in Th-232. Depending on the thorium layer's thickness and the moderating properties of the blanket, various correlations between the amounts of generated Pa-231, U-232 and U-233 can be obtained.

The results of analogous studies for uranium-fueled blankets of CTRs have been presented in [6,13]

3.2. URANIUM PROTECTION IN LWR TH-U FUEL

Calculations of the equilibrium uranium isotope vectors were carried out using cross-sections taken from the evaluated nuclear data files of the JENDL-3 library averaged over a neutron spectrum obtained from the GETERA code [7] for a typical VVER cell. The amount of the feed fuel required was derived from a condition that, at the end of cycle (fuel burn-up of 4% HM), the K_∞ of the core should equal that of a uranium fueled VVER. In this study, uranium protection of the feed fuel, ie., the U-232 fraction in the uranium of the feed fuel, was taken to be a variable parameter while the Pa-231 content in the feed fuel was kept constant.

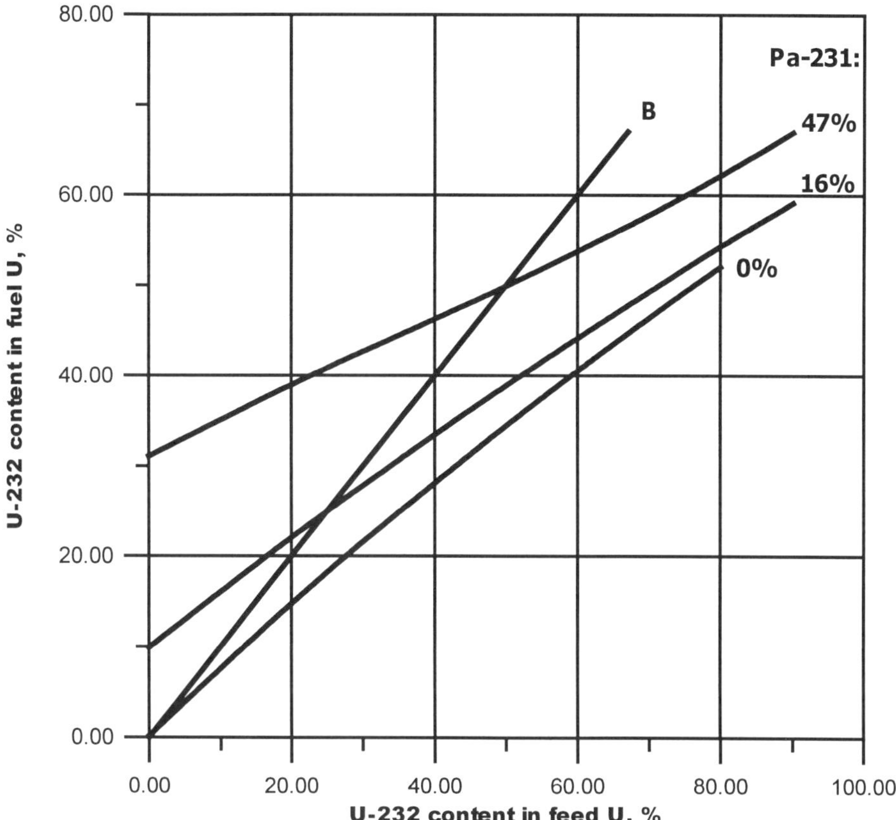

Fig. 2. Protection of the uranium in (discharged) fuel versus protection of the uranium in the feed fuel for various Pa-231 fractions in the feed fuel. "B" denotes the bisectrix.

The U-232 content of the uranium in the discharged fuel as a function of the U-232 content of the uranium in the feed fuel (for different Pa-231 loadings in the feed fuel) are shown in Fig. 2 for equilibrium multi-cycle operation. The curves in Fig. 2 demonstrate the correlation between the levels of uranium protection in the feed fuel when the feed contains Pa-231 in addition to uranium. Based on these results, it is possible to select an appropriate operational composition for the Th-U cycle. Hence, the inclusion of Pa-231 in the composition of the feed fuel makes it possible to attain an equal protection level for uranium in the discharged fuel and in the feed fuel.

The most distinctive characteristic of the curves plotted in Fig. 2 is the existence of a stationary irradiation composition that meets criticality requirements i.e. the case that the feed does not contain U-233, containing only U-232 and Pa-231 . This fact is explained by the good neutron multiplying properties of U-232 which offsets reactivity changes caused by fuel burn-up.

The decay heat generation of the uranium is considered a factor for its proliferation resistance. The rates of uranium decay heat generation for various feed fuel compositions are presented in Table 3. The rates of specific heat generation for weapons-grade plutonium (WGPu) and reactor-grade plutonium (RGPu) [1] are also presented in Table 3. It can be seen that denatured uranium fuel containing 18% U-232 is characterized by decay heat generation which exceeds that of RGPu by more than an order of magnitude, and exceeds by an even larger extent (ie., by 55 times) that of WGPu. This factor enhances protection of the U-233 in the uranium against utilization in nuclear explosive devices.

TABLE 3. Decay heat generation (q) and neutron emission by spontaneous fission (n_{sf}) in LWR fuel with equal uranium protection in the uranium of the discharged fuel and in the feed fuel.

Generation	WG Pu	RG Pu	U-232/U in both discharge and feed fuel (Pa-231/(Pa-231 + U) in feed)				
			5% (2%)	10% (5%)	18% (10%)	25% (16%)	50% (47%)
q^{U}, W/kg U	2.3	13	35	70	126	175	350
n_{sf}^{U}, 10^4 (n/sec)/kg U	6	38	0.0062	0.0121	0.0216	0.0298	0.0595
q^{fuel}, W/kg fuel	/ - /	/ - /	4.1	8.4	16	23	67
n_{sf}^{fuel}, 10^4 (n/sec)/kg fuel	/ - /	/ - /	0.0007	0.0014	0.0027	0.0039	0.0114
Feed ^{231}Pa/^{232}U/^{233}U, kg/(GWe*a)		/ - /	13 / 30 / 575	30 / 60 / 536	66 / 104 / 473	105 / 139 / 416	368 / 209 / 209

Values of specific heat generation in thorium fuel for the equilibrium cycle options analyzed are also shown in Table 3 also. One can see that U-233 denaturing with U-232 significantly increases the decay heat generation in thorium and, therefore, may need to be restricted by constraints imposed by permissible specific heat generation during processing of fuel and transportation. It may be concluded from Table 3 that the neutron emission rate of the denatured uranium is small, and therefore, cannot be considered to be a significant protection factor.

3.3 INCREASE OF FUEL BURN-UP FOR THE DENATURED FUEL CYCLE

Transition to denatured fuel containing isotopes of Pa-231, U-232, and U-233 will allow an increase in the fuel burn-up by many times as compared to standard fuel[14]. Some constraints may need to be taken into account at the fuel fabrication stage, as currently, the fuel rods for VVERs have enrichments of less than 4.4%.

With some simplifications, the chains for the isotopic transitions caused by neutron radiative capture reactions in the fuel may be presented as follows:

- uranium chain: $^{235}U \rightarrow {}^{236}U \rightarrow ...,$ $\quad ^{238}U \rightarrow {}^{239}Pu \rightarrow {}^{240}Pu \rightarrow ...$
- protactinium chain: $^{231}Pa \rightarrow {}^{232}U \rightarrow {}^{233}U \rightarrow {}^{234}U \rightarrow ...$

In the uranium chain, the initial and newly accumulated fissile isotopes (U-235 and Pu-239) are transformed into isotopes with poor fissile properties (U-236 and Pu-240, respectively) as a result of neutron radiative capture reactions. The principal distinction with the protactinium chain is that the U-232 isotope is a good fissile nuclide and, moreover, neutron radiative capture reactions on U-232 do not lead to the build-up of a neutron absorbing isotope, but leads to production of a fissile isotope, U-233, whose neutron multiplying properties are significantly better than those of U-232. This means that with burnup the content of an isotope with better multiplying properties (U-233) increases, and supports the reactor's criticality. In addition, the cross-section of Pa-231 for neutron radiative capture is significantly higher than that of U-238. Therefore, equal decreases in the amounts of these fertile isotopes with burnup also supports the reactor's criticality in the case that protactinium is included in the fuel.

The ideas explained above are confirmed by the data presented in Fig. 3. One can see that, if the required reactivity margin to compensate for neutron leakage and reactor control during operation corresponds to K∞=1.1, then a transition to denatured fuel will allow the fuel burn-up to be increased to near 40% HM. It is assumed that uranium containing 33% U-232 will meet the requirements for the proliferation resistance of the fuel.

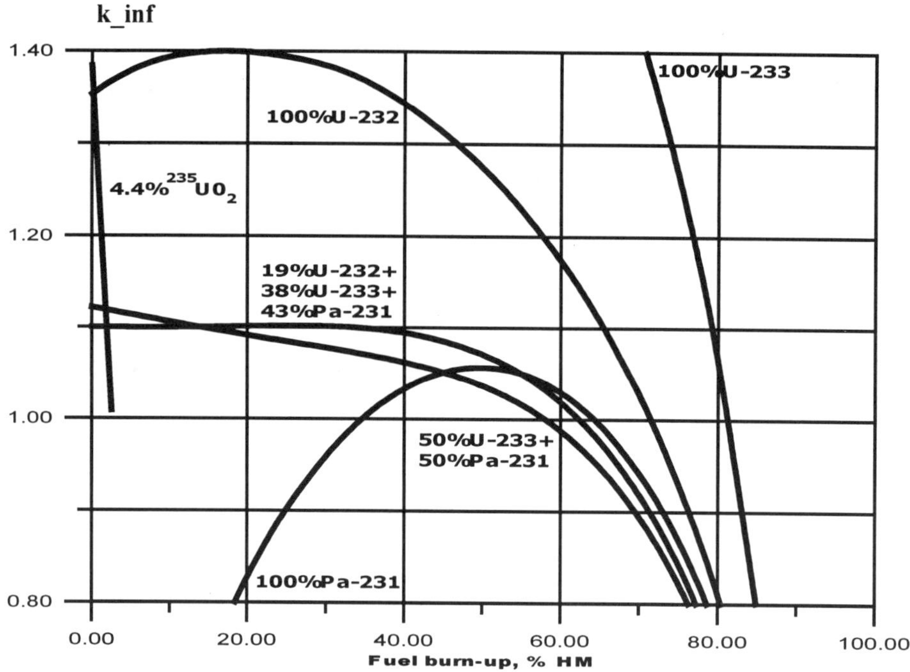

Fig. 3.   Dependence of K_∞ on fuel burn-up for various fuels.

It is also worth mentioning that there is a discrepancy in the multiplying properties of U-232 as given in the evaluated nuclear data files of the JENDL-3 and ENDF/B-VI librairies. The discrepancy is in the value for the fission neutron yield, ν_f for fissions caused by thermal neutrons. In JENDL-3, $\nu_f = 2.456$ while in ENDF/B-VI, $\nu_f = 3.13$. For the fuel cycles studied, U-232 plays an important role in maintaining reactor criticality. Therefore, the U-232 fission neutron yield value from the ENDF/B-VI library resulted in a significant increase in the potentially achievable fuel burn-up (to near 60% HM) and a decrease in the calculated U-233 critical content for achieving criticality compared to the corresponding values based on JENDL-3 .

It should also be noted that in these studies, one-group cross-sections obtained from the GETERA code for a VVER cell with Th-^{233}U fuel were used. In GETERA, the fission products are accumulated using an "effective" fission product model. The correctness of this model for high fuel burnups still needs to be confirmed and, therefore, results that depend on fuel burnup should be interpreted as estimated rather than exact values.

4. Conclusions

The proposed equilibrium fuel cycles, due to their special fuel compositions, exhibit advantages, such as, enhanced plutonium protection and significantly increased fuel burn-up (with only moderate changes in reactivity during operation).

5. Acknowledgment

The authors wish to express gratitude to Dr. L. Abagian for fruitful discussions concerning the nuclear data for the "exotic" nuclides of the fuel cycles analyzed.

References

1. Mark, J.C. (1993), Explosive Properties of Reactor-Grade Plutonium, Science & Global Security 4, p.111-128.
2. Rahn, F.J., Adamantiades, A.G., et al., (1984), A Guide to Nuclear Power Technology, A Wiley-Interscience Publication, John Wiley and Sons, New York.
3. Heising-Goodman, C.D. (1980), An Evaluation of the Plutonium Denaturing Concept as an Effective Safeguards Method", Nuclear Technology, 10, p.242-251.
4. Wydler, P., Heer, W., Stiller, P., and Wenger, H.U. (1980), A Uranium-Plutonium-Neptunium Fuel Cycle to Produce Isotopically Denatured Plutonium, Nuclear Technology, 6, p.115-120.
5. Ronen, Y., and Kimhi, Y. (1991), A Non-Proliferating Nuclear Fuel for Light Water Reactors, Nuclear Technology, 11, p.133-138.
6. Chmelev, A., Kryuchkov, E., Koulikov G., et al. (1996), A Conceptual Study of "Non-Proliferating" MOX Fuel Cycles for LWRs, Proceedings of the International Conference on the Physics of Reactors "PHYSOR'96", Japan, Vol.4, p.M-120-M-126.
7. Belousov, N., Bichkov, S., Marchuk, Y., et al. (1992), The Code GETERA for Cell and Polycell Calculations: Models and Capabilities, Proceedings of the 1992 Topical Meeting on Advances in Reactor Physics, March 8-11, 1992, Charleston, SC, USA, p.2-516-2-523.
8. Rolland-Piegue, C. (1995), Safeguards and Non-Proliferation for Advanced Fuel Cycles, IAEA Safeguards on Plutonium and HEU, Proceedings of the International Conference on Evaluation of Emerging Nuclear Fuel Cycle Systems "GLOBAL'95", September 11-14, 1995, Versailles, France, Vol.1, p.432-440.
9. Willrich, M., and Taylor, T.B. (1974), Nuclear Theft: Risks and Safeguards, Ballinger Publishing Company, Cambridge, Massachusetts, p.16-21.
10. Massey, J.V., and Schneider, A. (1982) The Role of Plutonium-238 in Nuclear Fuel Cycles, Nuclear Technology, 56, p.55-71.
11. Zaritskaya, T.S., Zaritsky, S.M., Kruglov, A.K., et al. (1980), Dependence of 232U Formation in Nuclear Reactors on Neutron Spectrum, Atomnaya Energiya, 48,

No.2, p.67-70.

12. Committee on International Security and Arms Control, National Academy of Sciences, Management and Disposition of Excess Weapons Plutonium, National Academy Press, Washington, D.C. (1994), p.151.

13. Gornostaev, B.D., Guriev, V.V., Orlov, V.V., Shatalov, G.E., et al. (1978), Experimental-Industrial Hybrid Fissile Materials Breeder, Proceedings of the 2nd Soviet-American Seminar on "Fusion-Fission", March 14 - April 1, 1977, Moscow, RRC-KI, Atomizdat, 1978, p.94-122.

14. Sinev, N.M., Baturov, B.B. (1984), Economics of Nuclear Industry, Moscow, Energoatomizdat.

DELAYED NEUTRON DATA FOR ACTINIDES OBTAINED FROM GLOBAL LEVEL MEASUREMENTS

T. A. PARISH
W. S. CHARLTON
Nuclear Engineering Department
Texas A&M University
College Station, TX 77843-3133 USA

1. Introduction

If weapons-grade plutonium is burned as fuel in LWRs, it may be necessary, for licensing purposes, to predict the behavior of full core loadings of such fuel during postulated accident transients. Operation with full core loadings of plutonium fuel will put increased emphasis on accurate knowledge of the "six group" delayed neutron parameters for plutonium and actinide isotopes. Comparisons of the six group delayed neutron data sets available in the literature for thermal neutron induced fission of Pu-239 indicate differences in the predicted reactor period for positive reactivity insertions of from 5 to 15 percent. It is desirable for reactor design and licensing analyses to decrease the range of uncertainty in the predicted reactor period. Therefore, improvements in the delayed neutron emission data for actinide isotopes is being sought. This paper summarizes some results based on recent measurements of delayed neutron emission from actinide isotopes conducted at Texas A&M University (TAMU).

Measurements of delayed neutron emission from various actinide isotopes have been performed at TAMU under the auspices of the Japan-United States Actinide Program. These measurements use a TRIGA fueled pool-type reactor as the neutron source. In general, delayed neutron data can be obtained and/or confirmed by three techniques. These are 1) measured at the "global" level by irradiating samples and measuring their subsequent neutron emission as a function of time, 2) calculated at the "individual" level by summing emission rates determined from (individual) fission product yield data and neutron emission probabilities, and 3) measured at the "integral" level by observing the transient response of reactors to known reactivity insertions.[1] "Global" level measurements have been recently performed at TAMU using 10 mg

T. A. Parish et al. (eds.),
Safety Issues Associated with Plutonium Involvement in the Nuclear Fuel Cycle, 213–224.
© 1999 *Kluwer Academic Publishers. Printed in the Netherlands.*

samples of U-235, Np-237 and Am-243. These measurments have focused on understanding the relationships between newly measured (global level) group parameters and the predicted six group parameters from the "individual" level approach, especially those that are included in ENDF/B-VI.

2. Experimental Procedure

Delayed neutron emission rates have been measured using the TAMU reactor to induce fission in actinide samples. The TAMU reactor core has a roughly rectangular shape and is made up of eighty-six TRIGA type fuel rods submerged in a water pool over 10.6 m (thirty-three feet) deep. Each TRIGA fuel rod has an active fuel length of 0.381 m (1.25 feet) and contains a mixture of uranium and zirconium hydride. As can be seen in the top view of the TAMU reactor core shown Fig. 1, the grid plate location at D-2 contains a pneumatic receiver that can be used for irradiating samples. The pneumatic receiver at D-2 has been used to investigate delayed neutron emission (global level measurements) from actinide samples in the work already reported by Saleh.[2] The neutron flux spectrum in the D-2 pneumatic location is well thermalized and a plot of the energy dependent flux at D-2 is presented in Fig. 2. The delayed neutron emission observed from the U-235 sample irradiated in the conventional pneumatic receiver at D-2 is predominately due to "thermal" neutron induced fission. However, for the Np-237 and Am-243 samples irradiated in the same location, the delayed neutron emission is predominately due to "fast" neutron induced fission.

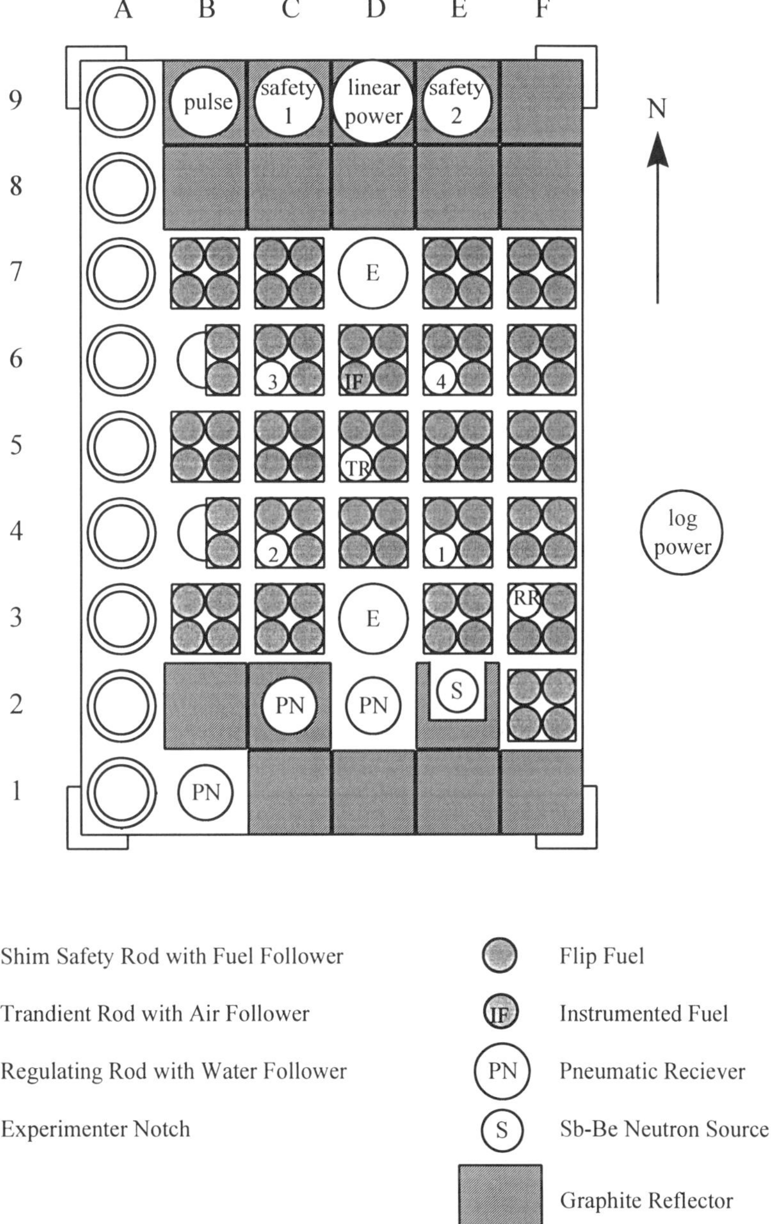

Fig.1. Top view of the Texas A&M University reactor core

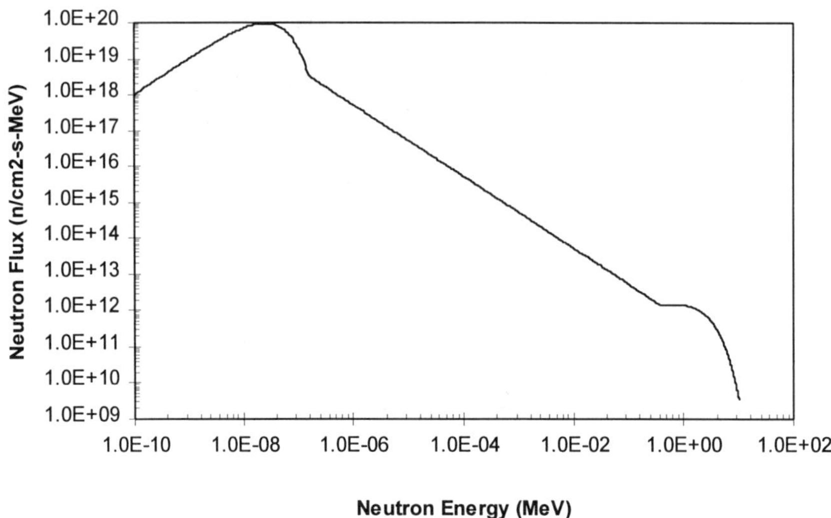

Fig. 2. Neutron flux as a function of energy in the pneumatic receiver at D-2

A pneumatic transfer system is used to position the actinide samples for irradiation. The pneumatic transfer system propels each sample first into the core, then to the detector array, and finally to the storage box as shown in Fig. 3. The tubing of the pneumatic system is made of polyethylene, and the samples are driven by carbon dioxide gas at a pressure of 0.69 MPa (100 psi). Since the pneumatic (sample) receiver is located inside the reactor core (bottom of pool) and the detector array is located at the reactor floor level (top of pool), the irradiation position to detector array distance shown in Fig. 3 is ~15 m (~50 feet). The sample transfer time from the irradiation position to the detector array has been measured using photosensors to be 0.55 seconds. This means that the shortest-lived delayed neutron group (i.e., group 6) can not be fully detected. Fortunately, group 6 accounts for less than 4% of the total delayed neutron production, and contributes less than 0.1% to the mean delayed neutron lifetime.

The delayed neutron emission from a sample is observed as a function of time using a BF_3 detector array as shown in Fig. 4. The BF_3 detectors are situated within a polyethylene cylinder. A lead collar surrounds the sample position to minimize the counts due to gamma ray pileup. The outside of the polyethylene cylinder is enclosed by a sheet of cadmium to maintain the background due to thermal neutrons as low as possible. The BF_3 detector array has been calibrated using a Cf-252 source of known strength and a Monte Carlo model of the detector array and polyethylene cylinder in order to compensate for the difference between the Cf-252 neutron energy spectrum and the energy spectrum of the delayed neutrons. The width of each time bin or "channel" of counting data is automatically adjusted using a computer so that the statistical error due to counting is maintained at a constant level of less than 3% in each time bin.

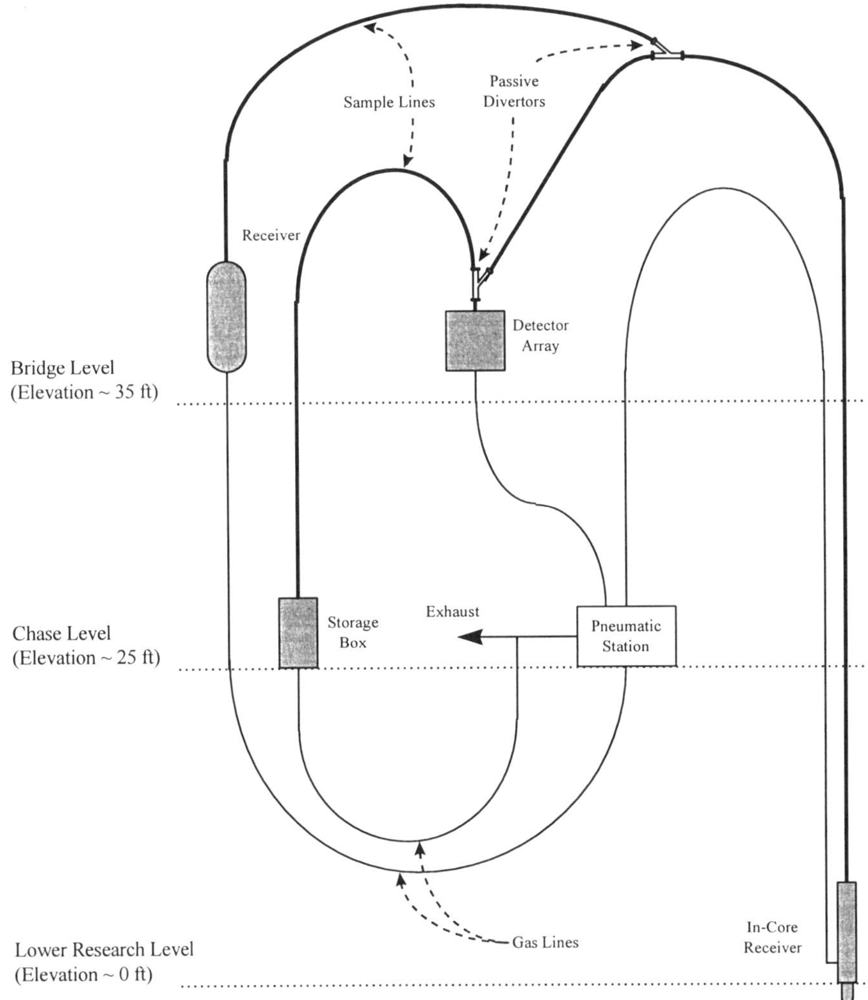

Fig. 3. Pneumatic transfer system paths at the TAMU reactor

The absolute (total) number of delayed neutrons emitted is found by integrating the observed net number of counts with corrections made for the detection efficiency and the counts due to the partially observable delayed group 6. The "group" decay constants and relative abundances are extracted directly from the observed counts versus time data by a nonlinear least squares fitting analysis. The fission rate in the sample needed to establish the value for the delayed neutron fraction is obtained by counting the samples on a well calibrated high purity germanium detector and observing the counts due to gamma rays from several prominent fission products.

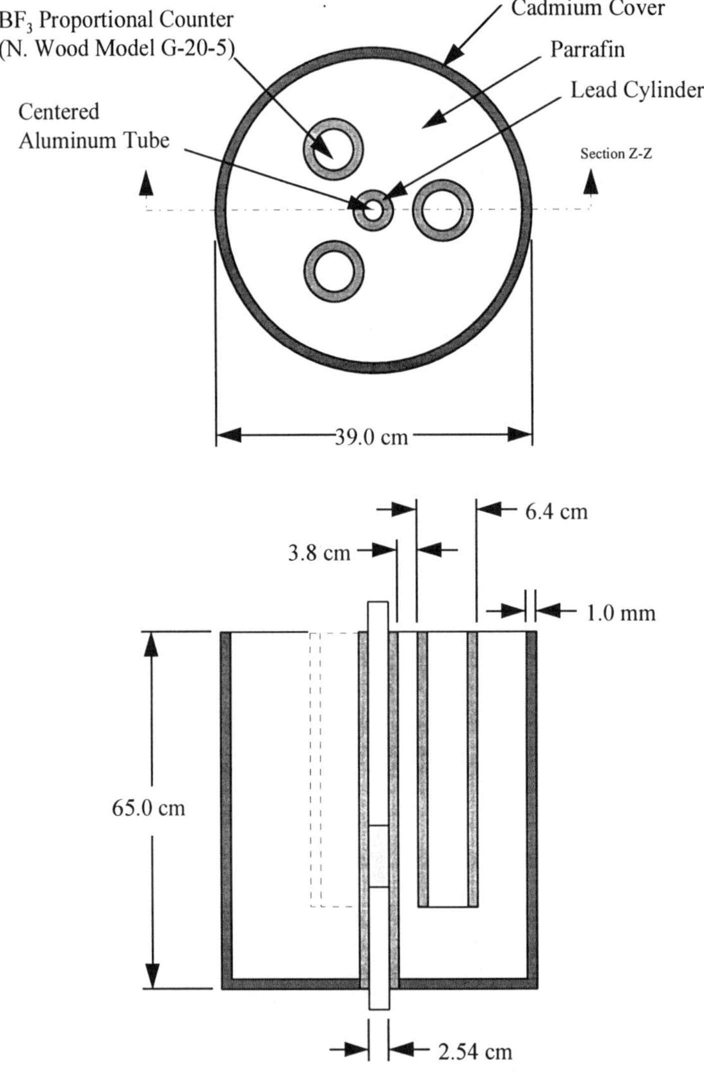

Fig. 4. BF3 detector array description

3. Results of the Delayed Neutron Measurements

Since Keepin's delayed neutron data were first reported, there has been much progress in the measurement of fission product yields and the properties of delayed neutron precursor isotopes. Extensive libraries have been compiled and six group delayed neutron parameter sets have been calculated, independent of experiments at the global level, using fission product yield data and delayed neutron emission probabilities for particular isotopes, i.e., the individual level approach.[3] Although early work, suggested a correspondence between the group parameters determined from global level experiments and the decay of particular precursor isotopes, Keepin's six group sets were obtained for the most part from fits to neutron emission versus time data with relatively little dependence on individual precursor information.

There are at least 271 individual delayed neutron precursors; however, less than 20 precursors account for over 80% of all delayed neutron emission. Table 1 identifies the longest-lived delayed neutron precursors based on U-235 thermal fission from the JEF2.2 library.[4] The delayed neutron precursors in Table 1 are arranged in order of decreasing half-life, and the Keepin group to which they might correspond is shown in the first column. The remaining values in Table 1 represent precursor labels, half-lives, decay constants, neutron emission probabilities, cumulative fission yields, the product of the neutron emission probability and fission yield and a calculated "group" average decay constant. Figure 5 displays a nomogram showing the contribution of the individual precursors (presented in Table 1) to the total delayed neutron emission following a thermal U-235 fission event as a function of precursor half life.

TABLE 1. Individual Delayed Neutron Precursors Corresponding to Keepin's Longest-Lived Groups

Group	Isotope	Half-life (sec)	P_n (%)	Yield (%)	P_n*Yield	Group Yield	Group λ (sec^{-1})
1	Br-87	55.69	2.52	2.0843	0.0525	*0.0525*	*0.0124*
2a	Cs-141	24.94	0.035	4.8032	0.00168	*0.2519*	*0.0283*
	I-137	24.5	7.14	3.5050	0.2503		
2b	Br-88	16.5	6.58	1.8206	0.1198	*0.1444*	*0.0416*
	Sb-134M	10.43	0.091	0.5154	0.000469		
	Te-136	17.5	1.3	1.8596	0.0242		
3	Se-87	5.6	0.36	0.7271	0.00262	*0.3338*	*0.1365*
	Br-89	4.38	13.8	1.4539	0.2006		
	Rb-92	4.51	0.010	4.8039	0.000514		
	Rb-93	5.7	1.35	3.5294	0.0476		
	I-138	6.41	5.46	1.4968	0.0817		
	As-84	5.5	0.28	0.2377	0.000666		
4	Rb-94	2.7	10.0	1.5884	0.1588	*0.5237*	*0.3254*
	Br-90	1.71	25.2	0.5379	0.1356		
	As-85	2.03	59.4	0.1606	0.0954		
	I-139	2.29	10.0	0.6269	0.0627		
	Y-97	1.21	0.08	2.8956	0.00232		
	Kr-93	1.29	1.95	0.5201	0.0101		
	Sb-135	1.7	17.6	0.1803	0.0317		
	Cs-143	1.77	1.62	1.6677	0.0270		

Again, the isotopes depicted are those that correlate to Keepin's first four (longest-lived) groups as shown in Table 1.

Some important suggestions are apparent from the previously presented data. First, since Br-87 is the longest-lived delayed neutron precursor, and it has a much longer half-life than any other precursor with a significant yield, macroscopic delayed neutron emission measurements for <u>any actinide</u> should lead to a decay constant for group 1 that is equal to that for Br-87, and the relative abundance value for group 1 from actinide to actinide should vary in accordance with the change in the fission yield for Br-87. Second, Keepin's group 2 is dominated by three precursor isotopes, I-137, Br-88, and Te-136 and, as can be seen in Fig. 5 and Table 1, the half-life of I-137 is significantly longer than those of Br-88 and Te-136. This would suggest that careful analysis of (global level) delayed neutron emission data might allow observation of these two components of Keepin's group 2. In Table 1, it is suggested that Keepin's group 2 be separated into two groups that are labeled 2a and 2b. Third, the individual fission product database may be valuable in fixing "group" decay constant values for at least the three or four longest-lived groups of Keepin's six groups. Recent data obtained from global level experiments support these suggestions.

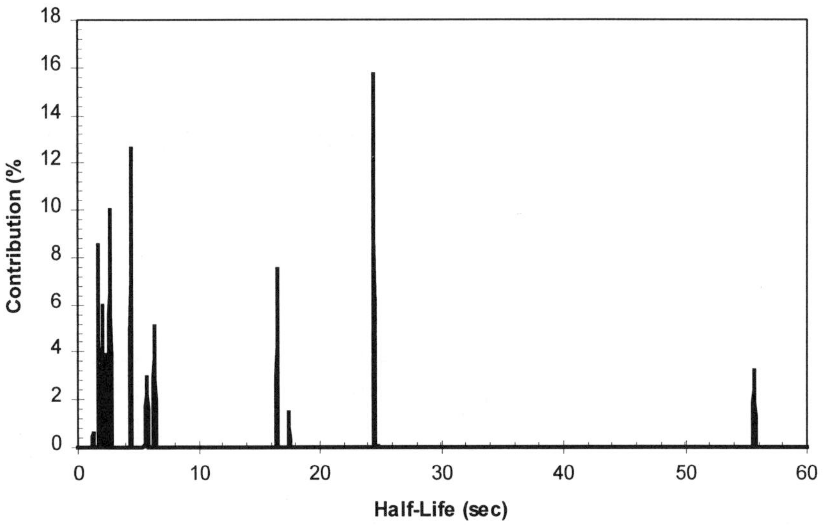

Fig. 5. Contributions of the twenty most important individual precursors to the total delayed neutron emission from U-235

Table 2 shows data by collected by Charlton that supports the first conclusion mentioned above. A detailed error propagation analysis has not yet been completed.

However, the results in Table 2 indicate that the "group 1" decay constants and relative abundances for U-235, Np-237, and Am-243 from the global level measurements consistently correspond to the individual isotope data for Br-87. Figs. 6a and 6b show count rate versus time data for fast fission of U-235. The data in Fig. 6a have already had the group 1 data stripped out. Fitting of the long-lived portion of this data from 200 to 300 seconds leads to a slope that yields a decay constant close to the value expected for I-137 (0.0283 sec^{-1}), i.e., the group 2a predicted in Table 1. When both the group 1 and group 2a data have been stripped, the count rate versus time data shown in Fig. 6b results. Fitting of the long-lived portion of this data, yields a decay constant close to the value (0.0416) anticipated for group 2b (corresponding to Br-88 and Te-136) as shown in Table 1. Comparisons of chi-squared for the fits using groups 2a and 2b as opposed to a single Keepin group 2 also clearly indicate improvement in the overall fit. This shows that the suggestions from the individual precursor data (for Br-87, I-137, Br-88, and Te-136) are in fact observable in the "global" measurements and that further effort to identify additional precursors and their relative abundances from the global measurements is warranted. The existence of groups 2a and 2b may have been unwittingly discovered by Waldo et al. in a series of experiments conducted at Lawrence Livermore National Laboratory.[5] From an analysis of Waldo's published data one finds that for seven of the fourteen isotopes studied (including U-233, Np-237, Pu-238, Pu-239, Pu-241, Am-242, Cm-245), Waldo had identified a group structure consistent with the subdivision of group 2 into a group 2a and a group 2b. Thus, it is prudent to assume that this higher order structure may be readily measurable in most actinides.

TABLE 2

Correspondence of the Fitted Decay Constants and Relative Abundances for the Longest Lived Delayed Neutron Group to Predicted Values for U-235, Np-237, and Am-243 (Based on Measured Data of Charlton [6])

Nuclide	Fit Decay Constant (sec^{-1})	Predicted Decay Constant (sec^{-1})	Exp. Relative Abundance	Predicted Relative Abundance
U-235	0.0125	0.0124	0.0327	0.033
Np-237	0.0124	0.0124	0.0313	0.030
Am-243	0.0124	0.0124	0.0167	0.0137

Finally, it is relatively easy to identify six or seven individual precursors to represent delayed groups 3 and 4. This warrants study, especially to determine if a universal set of decay constants might be specifiable for representing delayed neutron data for any actinide. It may also be possible to specify (single valued) "universal" decay constants for groups 5 and 6 even though they contain contributions from hundreds of precursors. This is due to the fact that, although delayed groups 5 and 6 account for

~15% of the delayed neutron emission, they contribute less than 1% to the mean delayed neutron lifetime, and therefore, are of relatively lesser importance in many reactor kinetics calculations.

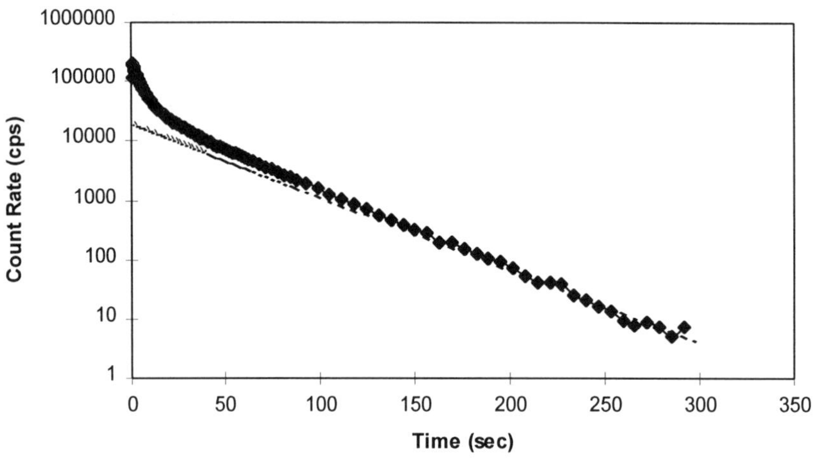

Fig. 6a. Measured neutron emission rate from U-235 with group 1 removed

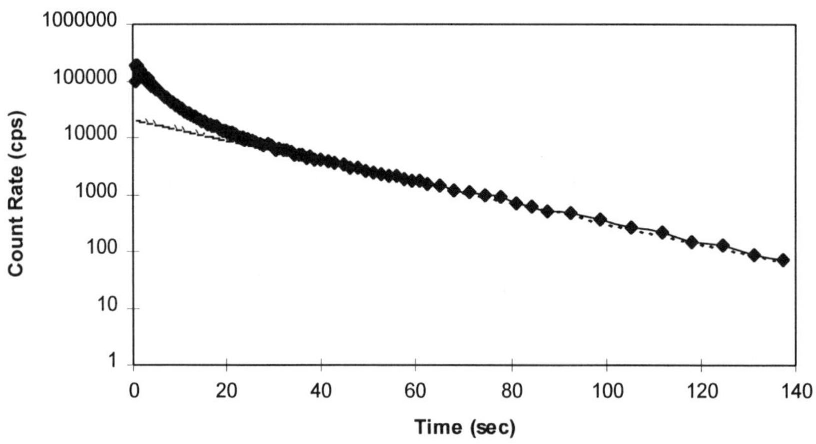

Fig. 6b. Measured neutron emission rate from U-235 with groups 1 and 2a removed

4. Conclusions

The major conclusions are (1) that much progress has been made in the application of individual fission product yield data for predicting delayed neutron emission and (2) that the individual and global approaches for obtaining delayed neutron group parameters need to be applied synergistically.

Careful fitting to measured neutron emission rate data for U-235, Np-237, and Am-243, guided by fission product yield data predictions, have verified that Keepin's delayed neutron group 1 corresponds closely to the properties of Br-87. Further, careful fitting has verified that Keepin's delayed neutron group 2 corresponds to three precursors, I-137, Br-88 and Te-136. The measured neutron emission rate data, if carefully fit, can be used to resolve Keepin's group 2 into two subgroups. The first subgroup corresponding to the decay of I-137 and the second subgroup corresponding to the decay of Br-88 and Te-136. The subgroups of Keepin's delayed neutron group 2 have been experimentally resolved in data from U-235, Np-237, and Am-243. The observed relative abundances appear to be in agreement with predictions based on the "individual level" approach for predicting delayed neutron emission.

Work is continuing to better determine the relationship between global neutron emission observations and "individual level" predictions. Data about the correspondence between eleven or twelve individual precursor's and Keepin's longest lived (four) groups may eventually allow replacement of Keepin's six arbitrary groups with eleven or twelve individual precursors and one or two "lumped" groups analogous to the manner in which fission products are treated in reactor physics calculations. This method may make possible (1) improved representations of delayed neutron data for a wide range of actinide isotopes and (2) more accurate reactor kinetics simulations.

5. Acknowledgements

This paper was based on research partially funded by the Japan/U.S. Actinides Program of ORNL in cooperation with sponsors at the Japan Atomic Energy Research Institute (JAERI). ORNL is operated by Martin Marietta Energy Systems, Inc., for the U. S. Department of Energy. The authors appreciate the interest in the development and improvement of nuclear data by S. Raman (ORNL) and T. Mukaiyama (JAERI).

6. Disclaimer

This paper was prepared as an account of work sponsored by an agency of the United States Government. Neither the United States Government nor any agency thereof, nor any of their employees, makes any warranty, express or implied, or assumes any legal liability or responsibility for the accuracy, completeness, or usefulness of any information, apparatus, product, or process disclosed, or represents that its use would not infringe privately owned rights. Reference herein to any specific commercial product, process, or service by trade name, trademark, manufacturer, or otherwise, does not necessarily constitute or imply its endorsement, recommendation, or favoring by the United States Government or any agency thereof. The views and opinions of authors expressed herein do not necessarily state or reflect those of the United States government or any agency thereof.

References

1. Blachot, J., Brady, M., Filip, A., Mills, R., and Weaver, D. (1990) Status of DelayedNeutron Data - 1990, Committee on Reactor Physics and Nuclear Data, Nuclear Energy Agency, Organization for Economic Cooperation and Development.
2. Saleh, H., Parish, T., Raman, S. and Shinohara, N. (1997) *Nuclear Science and Engineering* **125**, 51.
3. Brady, M. and England, T. (1989) *Nuclear Science and Engineering*, **103**, 129.
4. Blachot, J., Chung, C., Filip, A., and Storrer, F. (1995) JEF-2 Delayed Neutron Yield Data Evaluation for Emerging Fuel Cycle Systems, *International Conference on Evaluation of Emerging Nuclear Fuel Cycle Systems*, Versailles, France.
5. Waldo, R., Karam, R.and Meyer, R. (1981) *Physics Review C*, **23**, 1113.
6. Charlton W.(1997) Delayed Neutron Measurements from Fast Fission of Actinide Waste Isotopes," Masters Thesis, Texas A&M University.

BURNING OF PLUTONIUM IN ADVANCED MODULAR PEBBLE BED HTRs: THE MOST EFFECTIVE AND SAFE WAY FOR DISPOSITION

M. KHOROCHEV
E. TEUCHERT
H. RUETTEN
Institute for Safety Research and Reactor Technology
ISR KFA-Juelich GmbH
Postfach 1913
D-52425 Germany

Abstract

The consumption of reactor-grade and weapons-grade plutonium in a high temperature pebble bed reactor with a power of 350 MWt (PB-HTR-350) is described. It is shown that high temperature pebble bed reactors can be highly effective burners of plutonium. The study shows how pebble bed HTRs, operating with a Pu/Th or U/Pu fuel cycle using the two ball-type concept can provide a simple way of achieving very effective destruction of plutonium. With an initial loading of 3g of plutonium in each feed ball, one can achieve almost complete destruction of the Pu-239 and a substantial reduction in the total amount of plutonium (reduction to 5% of the amount loaded). This is possible because the plutonium and thorium (or uranium) can be inserted into different types of balls and each type of ball can be circulated through the reactor a different number of times to achieve the desired burnup.

1. Research Description

Compared to other reactors, (see Fig.1) high temperature reactors (HTRs) turn out to be the most suitable reactor type for the purpose of effectively destroying weapons-grade plutonium. In light water reactors (LWRs), the utilisation of plutonium is incomplete. Insertion of plutonium as a mixed oxide with uranium ($PuO2+UO2$) accomplishes only a partial burn up of the initial Pu as the consumption is, in part, compensated by the Pu

T. A. Parish et al. (eds.),
Safety Issues Associated with Plutonium Involvement in the Nuclear Fuel Cycle, 225–229.

© 1999 *Kluwer Academic Publishers. Printed in the Netherlands.*

build-up from neutron capture in U-238. The plutonium content of discharge LWR fuel is therefore about two-thirds of the original amount loaded. The fissile fraction of the discharged Pu is however reduced to less then 0.5. In LWRs, a considerably larger decrease in the amount of plutonium discharged can be achieved by using a mixed oxide fuel based on thorium rather than uranium. The Pu burnup in such a case is limited primarily by the need to maintain criticality. The HTR has some unique features which make it possible to achieve more complete burning of plutonium [1]. One of the main features is that different fuel isotopes can be inserted into different coated particles. In pebble bed reactors, the different coated particles can even be loaded into different balls, and the balls can be allowed to have different numbers of passes through the reactor until they reach the desired burn up.

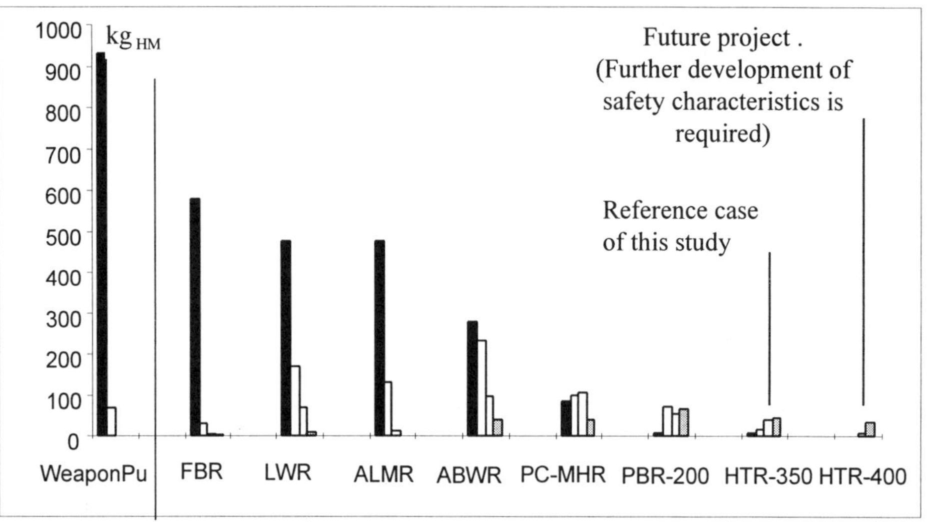

Fig. 1. Pu-vector at the end-of-life for fuel from different types of nuclear reactors loaded with 1 ton of weapons-grade Pu (930kg Pu-239 + 70kg Pu-240). From left to right, the unloaded Pu contents refer to the Pu-239, Pu-240, Pu-241 and Pu-242.

In addition, the technology for multi-layer coatings, is presently well developed. Multi-layer coatings improve the impermeability of the particles and reduce deleterious effects due to chemical interactions between the fission products and structural materials even at high burnups. Furthermore, if very high burnups are achieved, the problem of irradiated fuel storage may for practical purposes be significantly reduced.

The design for an advanced modular pebble bed reactor HTR-350 to burn plutonium (PB-HTR-350) has been carried out and the results are presented in this paper. This reactor design was developed from the conceptual design for the 200MWt-MODUL reactor (INTERATOM) while maintaining the same criteria for safety. Some basic features of the reactor are its power level of 350 MWt, its pre-stressed steel vessel and its annular core design. A parametric study to determine the plutonium consumption capability of the reactor was performed to find a design that satisfied the five goals listed in Table 1.

TABLE 1. Design criteria for the plutonium burning pebble bed reactor study

1	Maximum plutonium loading per GWd
2	Maximum in-situ plutonium burn up
3	Minimal Pu-content in the final fuel elements
4	Minimal U-235 requirement as driver material
5	Passive safety characteristics same as for the MODUL reactor concept

2. Advantageous Design Features of Pebble Bed HTRs

Pebble bed HTRs have several design features which make them particularly useful for plutonium destruction. Some of the important characteristics of pebble bed reactors, like the PB-HTR-350, specifically designed for plutonium burning are as follows:

- A pebble bed HTR operated with a Pu/Th or Pu/U fuel cycle provides one of the best ways to achieve a very high plutonium destruction rate. This is possible by inserting plutonium only into feed balls while thorium (or uranium) is inserted into separate breeding balls. The different ball types achieve final burnups of 800 and 120 MWd/kg_{HM}, respectively. This two ball type concept allows for maximum flexibility in meeting the requirements of the plutonium consumption mission.
- The choice of breeding material -- thorium for the reactor-grade plutonium case and uranium for the weapons-grade plutonium case -- is determined by safety criteria, (i.e. temperature coefficient of reactivity). The Doppler coefficient of Th-232 is less negative than that of U-238. In the weapons-grade plutonium case, in which the fraction of Pu-240 is only 7%, the negative contribution of Pu-240 to the moderator coefficient is not sufficient to obtain a negative total temperature coefficient. This leads to the choice of U-238 as the breeding (fertile) material, when weapons-grade plutonium is to be burned. In the reactor-grade plutonium case, use of thorium as the breeding material is feasible. Use of thorium as the breeding material excludes production of new plutonium, and the Pu burnup and U-235 consumption (reduced make-up to the driver fuel) are better than for the Pu/U-cycle. This favours the choice of the Pu/Th cycle for burning reactor-grade plutonium.
- In order to obtain rapid Pu-destruction, the lowest loading of heavy metal (HM) per breeding ball should be chosen. But, the safety criteria (temperature coefficients) lead to the selection of a higher HM loading (20g/ball) as the reference breeding ball HM loading. This loading yields acceptable temperature coefficients. Addition of fissile U-235 to the breeding balls is necessary to achieve high burnups. A further increase in

the HM loading in the breeding balls leads to poorer temperature coefficients due to the hardening of the neutron flux spectrum.

- The feed ball loading is determined by consideration of the plutonium mass balance and the temperature coefficient. Under these criteria, a loading of 3g(Pu)/ball has been found to be an optimal choice.

- By using different numbers of passes through the reactor for the feed and breeding balls, one can expose each ball type separately in the core until each type reaches its target burnup. It has been shown, that the lowest number of passes for the feed balls achieves the highest plutonium loading and the lowest uranium consumption. However, it also results in lower Pu-destruction and a less negative temperature coefficient.

- Variation of the fractions of the two types of balls leads to discovery of the optimal approach to satisfying the design criteria (see Table 1). The highest Pu consumption per GWd of energy release is achieved by using the highest fraction of feed balls. However, a rather large fraction of breeding balls has to be inserted in order to provide safety in the event of reactor cool down. The increase in reactivity due to reactor cool down and the required shutdown processes, ie., negative temperature coefficient, can be just achieved by using a .50 fraction of breeding balls.

A compromise solution considering all these tendencies brought about the selection of the reactor design parameters for a weapons-grade and reactor-grade plutonium burning PB-HTR-350 which are summarized in Table 2 and Fig.2.

TABLE 2. Performance of the reference designs for a Pu-loaded PB-HTR-350

Feed balls: Fissile material		Reactor-Pu	Weapons-Pu
Enrichment Pu-239 / Pu-240	%	70 / 30	93 / 7
Loading per fuel element	g/ball	3	3
Pu supply	kg/GWd	0.767	0.626
Pu content in the unloaded fuel	kg/GWd	0.082	0.068
In-situ -Burnup of Pu (total)	%	89.33	89.11
Breeding balls : Fertile material		Th-232	U-238
Make up - enrichment U-235/HM	%	8.5	10.0
Loading per fuel element	g/ball	20	20
Supply of U-235	Kg/GWd	0.297	0.417
Volume fractions of feed / breeding balls in the core		50 /50	50 / 50
Number of passes through the core feed /breeding balls		4 / 6	5 / 5
Total temperature coefficient	10^{-5}, $\Delta Keff/\Delta T$	-2.85	-5.99
Maximum temperature during LOCA	°C	1583	1576

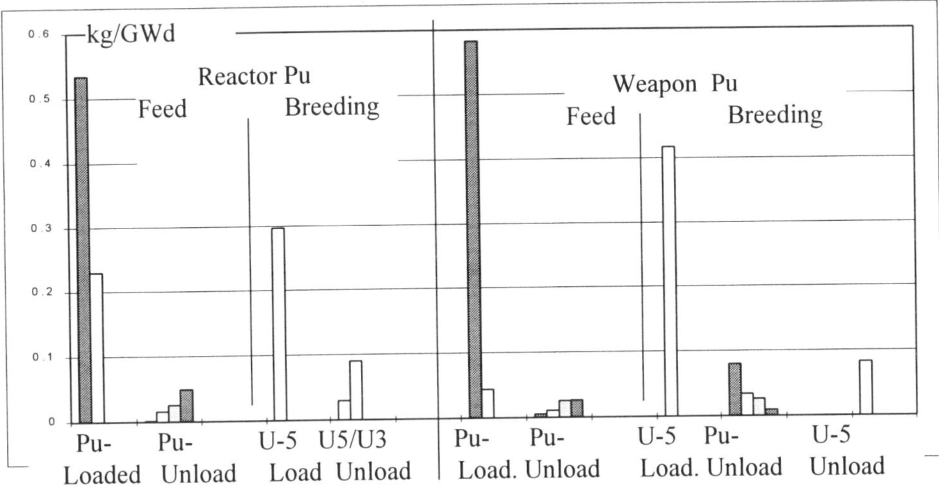

Fig. 2. Mass balance of fissile materials in feed and breeding balls for a PB-HTR-350 reactor burning reactor-grade and weapons-grade plutonium. From left to right, the Pu loaded and unloaded contents are Pu-239, Pu-240, Pu-241, Pu-242 and the U loaded and unloaded contents are U-235 and U-233.

3. Conclusions

Due to design conservatisms, this reactor could be constructed with minimal additional development. The available knowledge and technology based on experience at the AVR (which was operated for 21 years in Jülich Germany) has proven that such a reactor system is stable and safe under all conceivable scenarios, including nuclear, thermal, chemical and mechanical conditions.

Analyses of Loss of Coolant Accidents (LOCA) have been carried out for the different reference cases. These have shown that the passive removal of the decay heat is sufficiently high to keep the fuel temperature below 1600°C. At this temperature, the physical integrity of the coated particles remains uncompromised and the fission products are fully contained. The use of a helium turbine cycle for an HTR operating with Pu fuel elements, would ensure protection against accidental consequences, like those due to a water ingress accident.

Reference

1. Khorochev, M. (1997) Nutzung von Plutonium im Kugelhaufen-Hochtemperaturreaktor. Dissertation im Vorbereitung. (Usage of Plutonium in a Pebble Bed High Temperature Reactor, Ph.D. thesis to be published, KFA-Jüelich, Germany)

DENATURING EX-WEAPONS PLUTONIUM

A. G. TSIKUNOV
V. I. MATVEEV
V. A. CHERNY.
State Scientific Center of Russian Federation-Leipunsky Institute
of Physics and Power Engineering
Bondarenko sq.1
249020 Obninsk, Kaluga Region
Russian Federation

Abstract

The introduction of minor actinides (Np, Am) into fresh fuel will not only allow the efficient denaturing of weapons-grade plutonium, but will also allow the burning of waste minor actinides in the same nuclear fuel cycle. In this paper, the results of calculational studies on weapons-grade plutonium denaturation in a BN-800 type reactor are presented. Effective denaturation of a baseline amount (50 tons of ex-weapons Pu) with the addition of Np and Am into fresh MOX fuel can be carried out in one BN-800 type reactor over its 30-year lifetime. The required neptunium and americium (in addition to that existing in spent fuel of thermal reactors) can be obtained if reprocessing of the BN-800 type reactor spent fuel is carried out.

1. Introduction

The availability of weapons-grade plutonium as a result of the dismantling of nuclear weapons poses a new challenge, i.e. what is the best way to dispose of it? Since weapons-grade plutonium is the product of an expensive technology and has tremendous energy production potential, its utilization in nuclear reactors is worthy of study.

If the main criterion for disposition of weapons-grade plutonium is its isotopic degradation, then a case can be made that the most efficient means of disposition is through irradiation in thermal reactors. Unfortunately, very clearly defined criteria for

T. A. Parish et al. (eds.),
Safety Issues Associated with Plutonium Involvement in the Nuclear Fuel Cycle, 231–237.
© 1999 *Kluwer Academic Publishers. Printed in the Netherlands.*

"denaturing" are not available at present, although some physical prerequisites are evident, i.e. the greater the increase in Pu-240 and Pu-238 content, the more denaturated the original weapons-grade plutonium will be.

Table 1 presents typical compositions and characteristics of plutonium derived from reprocessing spent fuel from VVER-1000 and BN-800 reactor cores fuelled with MOX-fuel from weapons-grade plutonium.

These constitute two options for the disposition of weapons grade Pu. The utilization of weapons-grade (and reactor-grade plutonium) in thermal reactors because one observes:

- The production of minor actinides (Am, Cm) is increased relative to uranium fuel; and
- It is impossible to perform repeated recycles.

However, if fast reactors offer an advantage over thermal reactors on these points. In the case of fast reactors, enhanced denaturing of weapons-grade plutonium is possible due to:

- An increase in the fuel burn-up level; and
- The introduction of minor actinides (Np, Am) into the fresh fuel can increase Pu-238 production whose spontaneous emission rate of fission neutrons is 2.6 times and whose specific heat release is 80 times larger than that for Pu-240.

TABLE 1. Radiation characteristics and isotopic composition of weapons-grade Pu after irradiation in typical BN-800 and VVER-1000 reactors

Type of Pu	Weapons-grade Pu [1]	Pu from BN-800	Pu from VVER-1000
Isotope composition, % Pu-238	0.01	0.05	0.4
Pu-239	93.82	82.30	44.6
Pu-240	5.80	16.25	32.1
Pu-241	0.35	1.26	16.8
Pu-242	0.02	0.14	6.1
Heat release, W/g	2.26E-3	3.02E-3	7.0E-3
Spontaneous neutrons emission intensity, n/s·g	60.0	171	430

The first option (utilization of weapons-grade or reactor grade Pu in thermal reactors) leads to a decrease of weapons-grade plutonium consumption rate. The second scenario, in our view, will allow not only the performance of efficient more efficiently weapons-grade plutonium denaturing, but also will carry out the burning of minor actinides, that are presently wastes of the nuclear power cycle. In earlier publications [2,3,4] similar measures were proposed to enhance the protection of U-Pu fuel cycle of light-water reactors against proliferation of fissile materials. In these references, attention is centered on plutonium denaturing by the addition of Np-237 addition into the fuel. Therefore, when considering the large-scale denaturing of weapons-grade plutonium, the idea is to produce Np-237 in special thermonuclear installations [5]. In the case of using fast reactors to denature weapons-grade Pu, this special installation for Np-237 production are not needed. When solving the problem of weapons-grade plutonium denaturation in fast reactors, Am-241 is effectively interchangeable with Np-237 from a protection standpoint. The results of calculational studies on weapons-grade plutonium denaturing in a BN-800 type reactor (the second option) are presented below.

2. Weapons-Grade Plutonium Denaturing in BN-800 Type Reactors

In the calculational studies covering options with the addition of minor actinides (Np-237 and Am) into the fuel of a BN-800 type reactor, different amounts ranging from 1 to 5 % were considered. For americium, the following initial vector was considered (%):

Am-241/ Am-242m/ Am-243 = 63.7/ 2.1 / 34.2

Maximum MOX fuel burnup in the core was 10% HM. Two variants for minor actinides introduction into the fresh fuel were considered:

- minor actinides introduced into MOX fuel in the active part of a standard assembly; and
- minor actinides introduced into depleted uranium dioxide, located in the breeding part if a standard assembly. In the reactor, standard assembly design for the BN-800 reactor, only a lower breeding blanket of depleted uranium dioxide is provided.

In the latter case the influence of plutonium produced in axial breeding blanket on the plutonium final isotope composition was considered. This final isotopic composition was determined after joint chemical reprocessing of standard assembly fuel and breeding parts. Tables 2 and 3 present calculations which show the dependence of the downloaded plutonium isotope composition on the Np and Am content, respectively, in the "fresh" fuel.

234

TABLE 2. Dependence of the isotopic composition of Pu unloaded from the core and axial breeding blanket on Np-237 content in the fresh fuel

Np-237 content in fresh fuel, %		0	1	2	5
Isotope composition of Pu from standard assembly fuel parts, %	Pu-236	1.23E-6	5.96E-5	1.16E-4	2.78E-4
	Pu-238	5.14E-2	1.65	3.21	7.64
	Pu-239	82.3	80.9	79.6	75.9
	Pu-240	16.2	16.0	15.8	15.1
	Pu-241	1.26	1.25	1.23	1.18
	Pu-242	0.139	0.137	0.135	0.130
Isotope composition of Pu from standard assembly breeding parts, %	Pu-236	1.4E-7	4.84E-5	9.13E-5	1.95E-4
	Pu-238	1.86E-2	6.23	12.5	26.7
	Pu-239	95.3	89.0	83.5	70.0
	Pu-240	4.46	4.16	3.90	3.26
	Pu-241	0.185	0.173	0.161	0.135
	Pu-242	3.27E-3	3.05E-3	2.85E-3	2.38E-3

TABLE 3. Dependence of the isotopic composition of Pu, unloaded from the core and axial breeding blanket on the Am content in the fresh fuel

Np-237 content in fresh fuel, %		0	1	2	5
Isotope composition of Pu from standard assembly fuel parts, %	Pu-236	1.23E-6	1.31E-6	1.39E-6	1.62E-6
	Pu-238	5.14E-2	0.873	1.69	4.06
	Pu-239	82.3	81.4	80.5	77.8
	Pu-240	16.2	16.2	16.1	15.8
	Pu-241	1.26	1.25	1.25	1.22
	Pu-242	0.139	0.331	0.52	1.08
Isotope composition of Pu from breeding parts	Pu-236	1.4E-7	2.22E-7	3.0E-7	5.0E-7
	Pu-238	1.86E-2	3.36E	6.3	14.4
	Pu-239	95.3	91.3	87.5	77.5
	Pu-240	4.46	4.55	4.64	4.86
	Pu-241	0.185	0.18	0.174	0.16
	Pu-242	3.27E-3	0.71	1.37	3.13

The introduction of Np greatly influences the accumulation of Pu-236 and Pu-238 in the fuel and the introduction of Am influences the accumulation of Pu-238 and Pu-242, which are important for assessing the extent to which of weapons-grade plutonium is denatured.

Table 4 presents parameters (heat release and neutron emission intensity) of the plutonium produced after one cycle of irradiation for different initial contents of the minor actinides in the "fresh" fuel.

TABLE 4. Characteristics of weapons-grade Pu after one irradiation cycle in BN-800 type reactor with different minor actinide contents in the fresh fuel

			Spontaneous neutron emission intensity, n/g·s	Heat release, W/g
Initial weapons-grade Pu			60.0	2.26E-3
Weapons-grade Pu after one irradiation cycle in a thermal reactor			430	7.0E-3
Pu after one irradiation cycle in the core of a BN-800 core	Np	0.1%	171	3.02E-3
		1%	211	1.19E-2
		2%	250	2.06E-2
		5%	361	4.52E-2
Pu after one irradiation cycle in the axial zone of a BN-800	Np	0.1%	47.0	2.22E-3
		1%	208	3.69E-2
		2%	373	7.18E-2
		5%	744	1.51E-1
Pu after one cycle irradiation and mixing of the core and axial zone at introduction of Np only into the axial zone	Np	1%	174	5.05E-3
		2%	183	7.14E-3
		5%	205	1.19E-2
Pu after one cycle irradiation in the core	Am	1%	196	7.55E-3
		2%	220	1.21E-2
		5%	290	2.58E-2
Pu after one cycle irradiation in the axial zone	Am	1%	146	2.02E-2
		2%	240	3.73E-2
		5%	490	8.24E-2
Pu after one cycle irradiation and mixing of the core and axial zone at introduction of Am only into the axial zone	Am	1%	170	4.05E-3
		2%	175	5.07E-3
		5%	190	7.78E-3

Analysis of the data in Table 4 data shows that the addition of minor actinides into the "fresh" fuel can affect in a major fashion the denaturing characteristics of the fuel after one cycle. The addition of Np is the most efficient. Addition of approximately 2% minor actinides into MOX fuel (active part of standard assembly) or 5-6% into the depleted uranium part (axial breeding of the standard assembly) will result in denaturing parameters for the downloaded plutonium close to the parameters of plutonium obtained from thermal reactors.

Introduction of larger quantities of minor actinides into certain zones of a BN-800 type reactor could probably lead to the same characteristics of downloaded plutonium. These characteristics would preclude its use in military application. Besides, if the minor actinide content in the loaded fuel loaded to BN-800 type reactor (both in the core and in the breeding blankets) is $\geq 1\%$, long-lived minor actinide (Np-237, Am-241, Am-243) burning will occur. The extent of their burning during one cycle can be at a level of $\sim 30\%$ in the core, and a level of $\sim 20\%$ in the breeding zone.

3. Assessment of the Possibility of Carrying Out Large-Scale Denaturing of Weapons-Grade Plutonium in BN-800 Type Reactors

Addition of minor actinides into the "fresh" fuel leads to the deterioration of important safety parameters such as the sodium void reactivity effect (SVRE). Preliminary analysis has shown that for a BN-800 type reactor, if the core is totally submerged with sodium, the addition of up to 5-6 % minor actinides is possible without affecting much the SVRE [6]. This can be achieved through an increase in fuel enrichment, with a simultaneous decrease in the fuel fraction in the core. From the standpoint of the requirements on the SVRE values, limitations on the quantity of minor actinides introduced into the depleted uranium dioxide of the breeding zone are practically non-existant.

Is it possible to denature all weapons-grade plutonium available in Russia, using one BN-800 type reactor. From our analysis, 50 t of plutonium, a conservative quantity, would need approximately a total of 12 t of Np and Am to be denatured reliably.
This value is essentially the same for both the case of minor actinides introduction into the core fuel and for the case of minor actinides introduction into depleted uranium dioxide of the breeding blanket. According to our estimates, one can presently expect a total quantity of ~ 4 t of Np and Am in Russia. For the denaturing of 50 t weapons-grade plutonium, it would be necessary to carry out the reprocessing of spent fuel from a BN-800 type reactor with the aim of repeated entrainment of Np and Am into the cycle. In the case of multiple recycles, the total effective quantity of Np and Am, introduced into the fuel and the depleted UO_2 for plutonium denaturing in a BN-800 type reactor, is estimated in the following way:

$$P_{eff} = P_o \left[1 - (1 - q)^n \right] q^{-1}$$

where P_o - available quantity of Np and Am (4 t);

q - fraction of Np and Am burned in one BN-800 reactor during one cycle (in the core q=0.3; in the axial blanket q= 0.2);

n - number of fuel reprocessing cycles with high Np and Am recovery

The calculations show that for n $\geq$ 6 the effective Np and Am quantity (P_{eff}) becomes more than 12 MT. This result demonstrate it is possible to denature 50 MT of weapons-grade plutonium using only one BN-800 type reactor.

4. Conclusions

Effective denaturing of the baseline amount (50MT) of ex-weapons Pu by the addition of Np and Am into fresh MOX fuel can be carried out in one BN-800 type reactor over its 30-years life-time.

The required (in addition to existing quantities in spent fuel of thermal reactors) Np and Am needed for denaturing the baseline amount of ex-weapon Pu can be obtained through the reprocessing of BN-800 type reactor spent fuel reprocessing and through the repeated inclusion of "non-burned" Np and Am into the fuel.

References

1. D. Berwald, and A. Favale. (23-27 May 1994), Potential Role of Emerging Accelerator - Based Transmutation Technology in U.S. Plutonium and High - Level Waste Deposition (Northrop - Grumman Corporation, Bethpage, NY).- The Second International Seminar on Transmutation of Long - lived Radioactive Wastes and Conversion of Weapons - Grade Plutonium Based on the Use of Proton Accelerators, ITEPH, Moscow, P.71-84.

2. Carolyn D. Heising-Goodman.(Oct. 1980), An Evaluation of the Plutonium Denaturing Concept as an Effective Safeguards Method.- Nuclear Technology, Vol.50, -P.242-251.

3. P. Wydler, W. Heer, P. Stiller, and H. U. Wenger. (June 1980), A Uranium-Plutonium Fuel Cycle to Produce Isotopically Denatured Plutonium.-Nuclear Technology,Vol.49,-P.115-120.

4. Yigal Ronen, and Yehoshua Kimhi. (Nov.1991), A "Nonproliferating" Nuclear Fuel for Light Water Reactors.- Nuclear Technology, Vol.96, P.133-138.

5. A. N. Chmelev, E. F. Kryuchkov, G. G. Koulikov, V. A. Apse, A. F. Lashin, Nuclear Power Industry, No. 4, pp. 34-40.

6. V.A. Yeleseev, I. Yu. Krivitsky, V. I. Matveev, Efficiency of plutonium burning in BN-800 and BN-600 reactor cores. Preprint IPPE-2598, Obninsk.

AUTHOR AND SUBJECT INDEX

T. A. Parish et al. (eds.),
Safety Issues Associated with Plutonium Involvement in the Nuclear Fuel Cycle, 239–241.
© 1999 *Kluwer Academic Publishers. Printed in the Netherlands.*